International
Association
of Fire Chiefs

Hazardous Materials

Awareness and Operations

Student Workbook

JONES AND BARTLETT PUBLISHERS
Sudbury, Massachusetts
BOSTON TORONTO LONDON SINGAPORE

Jones and Bartlett Publishers
World Headquarters
40 Tall Pine Drive
Sudbury, MA 01776
978-443-5000
www.jbpub.com

National Fire Protection Association
1 Batterymarch Park
Quincy, MA 02169
www.NFPA.org

International Association of Fire Chiefs
4025 Fair Ridge Drive
Fairfax, VA 22033
www.IAFC.org

Jones and Bartlett Publishers Canada
6339 Ormindale Way
Mississauga, Ontario L5V 1J2
Canada

Jones and Bartlett Publishers International
Barb House, Barb Mews
London W6 7PA
United Kingdom

Jones and Bartlett's books and products are available through most bookstores and online booksellers. To contact Jones and Bartlett Publishers directly, call 800-832-0034, fax 978-443-8000, or visit our website www.jbpub.com.

Substantial discounts on bulk quantities of Jones and Bartlett's publications are available to corporations, professional associations, and other qualified organizations. For details and specific discount information, contact the special sales department at Jones and Bartlett via the above contact information or send an email to specialsales@jbpub.com.

Editorial Credits

Author: Leo J. DeBobes, MA (OS&H), CSP, CHCM, CSC, CPEA, EMT-D

Production Credits

Chief Executive Officer: Clayton E. Jones
Chief Operating Officer: Donald W. Jones, Jr.
President, Higher Education and Professional Publishing: Robert W. Holland, Jr.
V.P., Production and Design: Anne Spencer
V.P., Manufacturing and Inventory Control: Therese Connell
Publisher, Public Safety: Kimberly Brophy
Senior Acquisitions Editor—Fire: William Larkin
Managing Editor: Carol Guerrero
Associate Editor: Laura Burns
Production Manager: Jenny Corriveau
Production Assistant: Tina Chen
Photo Research Manager/Photographer: Kimberly Potvin
Assistant Photo Researcher: Meghan Hayes
Director of Sales—Public Safety: Matthew Maniscalco
Director of Marketing: Alisha Weisman
Marketing Manager—Fire: Brian Rooney
Text Design: Anne Spencer
Cover Design: Kristin E. Parker
Composition: diacriTech
Cover Image: © Chris Landsberger, Topeka Capital Journal/AP Photos
Text Printing and Binding: Courier Kendallville
Cover Printing: Courier Kendallville

ISBN: 978-0-7637-7121-8

The procedures and protocols in this book are based on the most current recommendations of responsible sources. The International Association of Fire Chiefs (IAFC), National Fire Protection Association (NFPA®), and the publisher, however, make no guarantee as to, and assume no responsibility for, the correctness, sufficiency, or completeness of such information or recommendations. Other or additional safety measures may be required under particular circumstances.

Additional photographic and illustration credits appear on page 171, which constitutes a continuation of the copyright page.

Notice: The individuals described in "Fire Alarms" throughout this text are fictitious.

6048

Printed in the United States of America
13 12 11 10 9 8 7 6 5 4 3 2

Contents

Answer Key

Student Resources

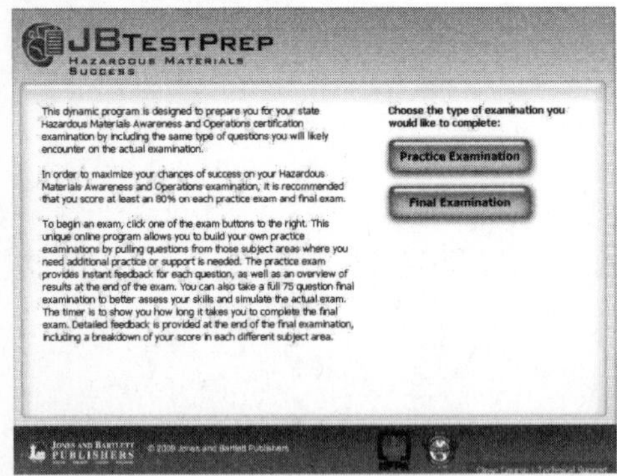

JB Test Prep: Hazardous Materials Success

ISBN-13: 978-0-7637-7122-5

JB Test Prep: Hazardous Materials Success is a dynamic program designed to prepare students to sit for Hazardous Materials Awareness- and Operations-level certification examinations by including the same type of questions they will likely see on the actual examination.

It provides a series of self-study modules, organized by chapter and level, offering practice examinations and simulated certification examinations using multiple-choice questions. All questions are page referenced to *Hazardous Materials Awareness and Operations* for remediation to help students hone their knowledge of the subject matter.

Students can begin the task of studying for Hazardous Materials Awareness- and Operations-level certification examinations by concentrating on those subject areas where they need the most help. Upon completion, students will feel confident and prepared to complete the final step in the certification process—passing the examination.

Hazardous Materials Awareness and Operations Field Guide

ISBN-13: 978-0-7637-7701-2

A quick reference to essential information for hazardous materials responders, *Hazardous Materials Awareness and Operations Field Guide* includes charts and tables to provide easy access to key topics. Designed to withstand the elements, this field guide is pocket-sized, spiral bound, and water resistant.

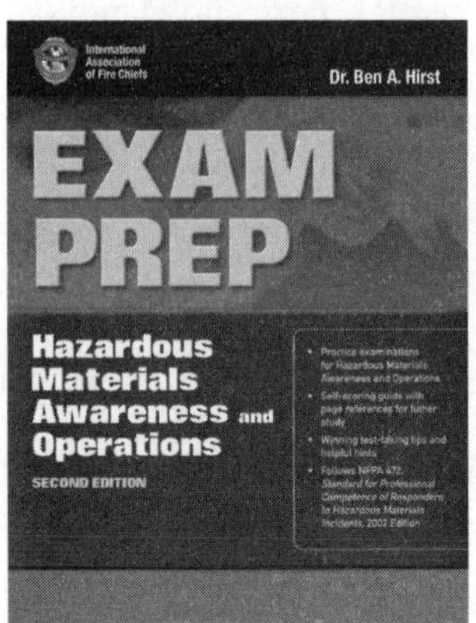

Exam Prep: Hazardous Materials Awareness and Operations, Second Edition

ISBN-13: 978-0-7637-5838-7

The second edition of *Exam Prep: Hazardous Materials Awareness and Operations* is designed to thoroughly prepare you for a Hazardous Materials certification, promotion, or training examination by including the same type of multiple-choice questions you are likely to encounter on the actual examination. To help improve examination scores, this preparation guide follows Performance Training Systems, Inc.'s Systematic Approach to Examination Preparation®. *Exam Prep: Hazardous Materials Awareness and Operations* is written by fire personnel explicitly for fire personnel, and all content has been verified with the latest reference materials and by a technical review committee.

Benefits of the Systematic Approach to Examination Preparation include:

- Emphasizing areas of weakness
- Providing immediate feedback
- Learning material through context and association

Exam Prep: Hazardous Materials Awareness and Operations, Second Edition includes:

- Practice examinations for Hazardous Materials Awareness and Operations levels
- Self-scoring guide with page references for further study
- Winning test-taking tips and helpful hints
- Coverage of NFPA 472, *Standard for Competence of Responders to Hazardous Materials/Weapons of Mass Destruction Incidents*, 2008 Edition

Technology Resources

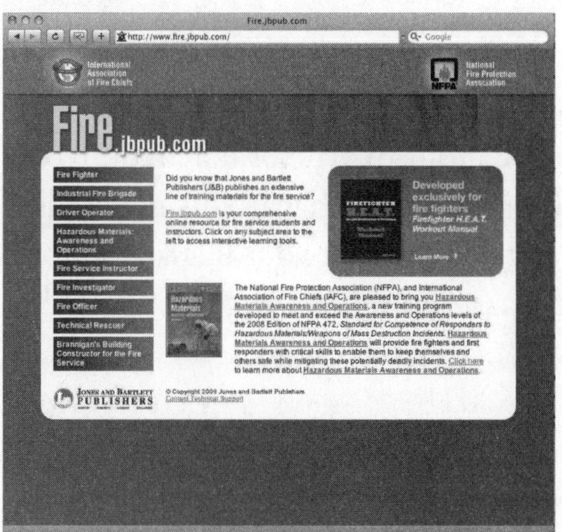

www.Fire.jbpub.com

This site has been specifically designed to complement *Hazardous Materials Awareness and Operations* and is regularly updated. Resources available include:

- **Chapter Pretests** that prepare students for training. Each chapter has a pretest and provides instant results, feedback on incorrect answers, and page references for further study.
- **Interactivities** that allow students to reinforce their understanding of the most important concepts in each chapter.
- **Hot Term Explorer**, a virtual dictionary that allows students to review key terms, test their knowledge of key terms through quizzes and flashcards, and complete crossword puzzles.

HazMat Online

ISBN-13: 978-0-7637-3476-3

HazMat Online is a series of online modules that brings Hazardous Materials Awareness and Operations level material to life with interactive audio, video, Flash™ animation, photographs, illustrations, and case-based scenarios. *HazMat Online* is designed to enhance students' learning experience by allowing them to experiment with key concepts and skills in the safety of a virtual environment.

Students progress through the didactic portion of each module online, followed by an interactive review of the material to evaluate retention of key concepts presented. *HazMat Online* can be used as:

- Annual refresher training for students who have taken the Hazardous Materials Operations Level 472 course
- Distance learning course
- A review tool to assess students' progress and clarify more challenging subjects

Hazardous Materials: Overview

Workbook Activities

The following activities have been designed to help you. Your instructor may require you to complete some or all of these activities as a regular part of your hazardous materials training program. You are encouraged to complete any activity that your instructor does not assign as a way to enhance your learning in the classroom.

Chapter Review

The following exercises provide an opportunity to refresh your knowledge of this chapter.

Matching

Match each of the terms in the left column to the appropriate definition in the right column.

______	**1.** Awareness level	**A.**	Training that provides the ability to enter heavily contaminated areas using the highest levels of protection
______	**2.** Technician level	**B.**	The impure substance left after manufacturing
______	**3.** Hazardous material	**C.**	A federal agency that ensures safe manufacturing, use, transportation, and disposal of hazardous substances
______	**4.** EPA	**D.**	A facility that presents a high potential for loss of life
______	**5.** Hazardous waste	**E.**	The body that develops and maintains nationally recognized minimum consensus standards on many areas of fire safety and hazardous materials
______	**6.** NFPA	**F.**	Training that provides the ability to recognize a potential hazardous emergency and isolate the area
______	**7.** Target hazard	**G.**	Any material that poses an unreasonable risk of damage or injury to persons, property, or the environment if not properly controlled
______	**8.** SARA	**H.**	Liaison between local and state levels of emergency response authorities
______	**9.** SERC	**I.**	A detailed profile of a single chemical or mixture of chemicals provided by the manufacturer or supplier of a chemical
______	**10.** MSDS	**J.**	Law that affects how fire departments respond in a hazardous materials emergency

Multiple Choice

Read each item carefully, and then select the best response.

______ **1.** A material that poses an unreasonable risk to the health and safety of the public and/or the environment if it is not controlled properly during handling, processing, and disposal is called a:

A. hazardous waste.
B. hazardous material.
C. hazardous target.
D. hazardous substance.

CHAPTER

1

______ 2. Which of the following is a nongovernment agency that issues fire response standards?
A. CANUTEC
B. NFPA
C. OSHA
D. EPA

______ 3. In the United States, the federal document containing the hazardous materials response competencies is known as:
A. NFPA.
B. EPCRA.
C. SARA.
D. HAZWOPER.

______ 4. In the United States, which federal government agency enforces and publicizes laws and regulations that govern the transportation of goods by highways, rail, and air?
A. Environmental Protection Agency (EPA)
B. Occupational Safety and Health Administration (OSHA)
C. State Emergency Response Commission (SERC)
D. Department of Transportation (DOT)

______ 5. What act requires a business that handles chemicals to report storage type, quantity, and storage methods to the fire department and the local emergency planning committee?
A. Superfund Amendments and Reauthorization Act
B. Local Emergency Planning Committee Act
C. Emergency Planning and Community Right to Know Act
D. Occupational Safety and Health Act

______ 6. Which of the following is a group that gathers information about hazardous materials and disseminates that information to the public?
A. LEPC
B. NFPA
C. SARA
D. EPA

______ 7. Which of the following is a state group that acts as a liaison between local- and state-level response authorities?
A. SARA
B. EPA
C. SERC
D. MSDS

______ 8. Which response level is trained to recognize a hazardous materials emergency and call for assistance?
A. Awareness
B. Operations
C. Technician
D. Specialist

______ 9. Which response level is trained to take defensive actions?
A. Awareness
B. Operations
C. Technician
D. Specialist

_____ **10.** Which response level is trained to take offensive actions?
- **A.** Awareness
- **B.** Operations
- **C.** Technician
- **D.** Specialist

Vocabulary

Define the following terms using the space provided.

1. Local emergency planning committee (LEPC):

____

____

____

2. Material safety data sheet (MSDS):

____

____

____

3. Specialist level:

____

____

____

4. HAZWOPER:

____

____

____

5. Operations level:

____

____

____

Fill-in

Read each item carefully, and then complete the statement by filling in the missing word(s).

1. The ______________ ______________ ______________ regulates and governs issues relating to hazardous materials in the environment.

2. A(n) ______________ ______________ ______________ ______________ is a detailed profile of a single chemical, or mixture of chemicals, provided by the manufacturer and/or supplier of a chemical.

3. Hazardous materials incidents are ______________ complicated than most structural firefighting incidents.

4. ______________ activities enable agencies to develop logical and appropriate response procedures for anticipated incidents.

5. Awareness-level skills are ____________________ and not defensive.

6. Hazardous materials ____________________ take offensive actions.

7. ____________________ is the *Standard for Competence of Responders to Hazardous Materials/Weapons of Mass Destruction Incidents.*

8. The federal document containing the hazardous materials response competencies is known as ____________________.

9. _______________ _______________ is the material that remains after a manufacturing plant has used some chemicals, and they are no longer pure.

10. States that have adopted OSHA safety and health regulations are called _______________-_______________states.

True/False

If you believe the statement to be more true than false, write the letter "T" in the space provided. If you believe the statement to be more false than true, write the letter "F."

1. ______ The ability to recognize a potential hazardous materials incident is critical to ensuring one's safety.
2. ______ The Emergency Planning and Community Right to Know Act was one of the first laws to affect how fire departments respond in a hazardous materials emergency.
3. ______ Each state has a State Emergency Response Commission (SERC) that acts as a liaison between local and state levels of authority.
4. ______ The actions taken at hazardous materials incidents are largely dictated by the chemicals involved.
5. ______ Fires require a less straightforward response than do hazardous materials incidents.
6. ______ When approaching a hazardous materials event, you should make a conscious effort to change your response perspective.
7. ______ Response agencies should not preplan target hazards owing to the health issues involved in such planning.
8. ______ The goal of a fire fighter is to favorably change the outcome of a hazardous materials incident.
9. ______ The SARA regulates and governs issues relating to hazardous materials and the environment.
10. ______ The EPA's version of HAZWOPER is in Title 40, *Protection of the Environment, Part 311, Worker Safety.*

Short Answer

Complete this section with short written answers using the space provided.

1. Identify the four levels of hazardous materials training and competencies, according to NFPA 472.

____

____

____

____

2. Discuss the Superfund Amendments and Reauthorization Act.

____

____

____

____

CLUES

Word Fun

The following crossword puzzle is an activity provided to reinforce correct spelling and understanding of terminology associated with hazardous materials. Use the clues provided to complete the puzzle.

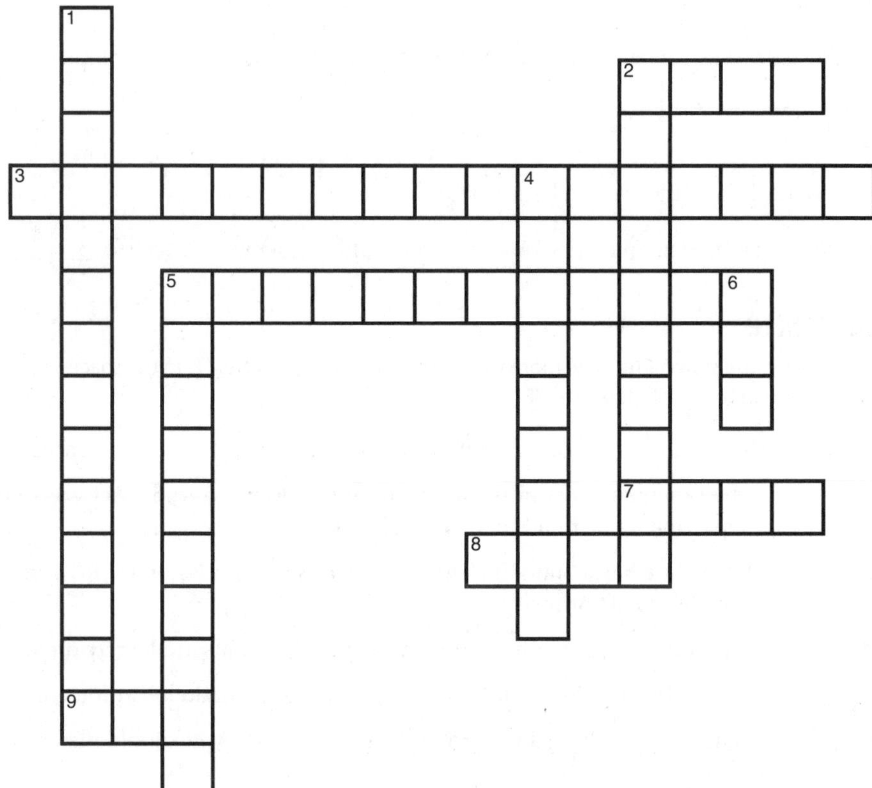

Across

2 The federal agency that regulates worker safety and, in some cases, responder safety. (abbreviation)

3 Any material that poses an unreasonable risk of damage or injury to persons, property, or the environment if it is not properly controlled. (2 words)

5 A facility that presents a high potential for loss of life or serious impact to the community resulting from fire, explosion, or chemical release. (2 words)

7 The association that maintains nationally recognized minimum consensus standards on many areas of fire safety. (abbreviation)

8 A detailed profile of a chemical, provided by the manufacturer and/or supplier of that chemical. (abbreviation)

9 The federal agency that ensures safe manufacturing, use, transportation, and disposal of hazardous materials. (abbreviation)

Down

1 A substance that remains after a process or manufacturing plant has used some of the material, and it is no longer pure. (2 words)

2 The level of training that should allow responders to be able to recognize and isolate hazardous materials and deny entry to other responders.

4 The level of training that provides first responders with the ability to recognize a potential hazardous materials emergency.

5 Responders trained to this level are able to enter heavily contaminated areas and take offensive actions.

6 The federal department that publicizes and enforces rules and regulations that relate to the transportation of many hazardous materials. (abbreviation)

Fire Alarms

The following real case scenarios will give you an opportunity to explore the concerns associated with hazardous materials. Read each scenario, and then answer each question in detail.

1. You are participating in a hazardous materials awareness class, and the instructor asks you to define the term "hazardous materials." What is your response?

____
____
____
____
____
____

2. You have just completed your hazardous materials training. Your Lieutenant gives you an assignment to complete a mock preincident plan for a hazardous materials response to an occupancy near your station. How should you proceed?

____
____
____
____
____
____

Hazardous Materials: Properties and Effects

Workbook Activities

The following activities have been designed to help you. Your instructor may require you to complete some or all of these activities as a regular part of your hazardous materials training program. You are encouraged to complete any activity that your instructor does not assign as a way to enhance your learning in the classroom.

Chapter Review

The following exercises provide an opportunity to refresh your knowledge of this chapter.

Matching

Match each of the terms in the left column to the appropriate definition in the right column.

______	**1.** Vapor	**A.**	The ability of a material to cause damage upon skin contact
______	**2.** Lower explosive limit	**B.**	An expression of a fuel/air mixture, defined by upper and lower limits, that reflects the concentration of flammable vapor in air that is necessary for a combustible material to burn properly
______	**3.** Corrosivity	**C.**	Fluid build-up in the lungs
______	**4.** Toxicology	**D.**	The weight of a gas as compared to an equal volume of dry air
______	**5.** Flammable range	**E.**	The gas phase of a substance
______	**6.** Vapor density	**F.**	The minimum amount of gaseous fuel that must be present in the air mixture for the mixture to be flammable or explosive
______	**7.** Pulmonary edema	**G.**	A cancer-causing agent
______	**8.** Vapor pressure	**H.**	The study of the adverse effects of chemical or physical agents on living organisms
______	**9.** Carcinogen	**I.**	The pressure exerted by a liquid's vapor, inside a closed container, when liquid and vapor phases are in equilibrium
______	**10.** Expansion ratio	**J.**	A description of a volume increase that occurs when a liquid changes to a gas

Multiple Choice

Read each item carefully, and then select the best response.

______ **1.** The characteristics of a chemical that are measurable are:

A. physical properties and chemical properties.
B. chemical properties.
C. states of matter.
D. radiation agents.

CHAPTER 2

______ **2.** The first step in understanding the hazard of any chemical involves identifying:
A. physical properties.
B. chemical properties.
C. the state of matter.
D. radiation agents.

______ **3.** The expansion ratio is a description of the volume increase that occurs when a material changes from:
A. a liquid to a solid.
B. a solid to a gas.
C. a solid to a liquid.
D. a liquid to a gas.

______ **4.** The ability of a chemical to undergo a change in its chemical make-up, usually with a release of some form of energy, is called:
A. a property change.
B. a physical change.
C. chemical reactivity.
D. a change of state.

______ **5.** Steel rusting and wood burning are examples of:
A. physical changes.
B. chemical changes.
C. vaporization.
D. ionization.

______ **6.** The temperature at which a liquid will continually give off vapors, and will eventually turn completely into a gas, is called the:
A. flash point.
B. vaporization point.
C. boiling point.
D. gas point.

______ **7.** The weight of an airborne concentration as compared to an equal volume of dry air is the:
A. vapor density.
B. vapor ratio.
C. flammable range.
D. explosive ratio.

______ **8.** Air has a set vapor density value of:
A. 0.59.
B. 1.0.
C. 2.4.
D. 3.8.

______ **9.** The vapor pressure at the standard atmospheric temperature of 68°F (20°C) can be expressed in pounds per square inch, atmospheres, and millimeters of mercury, as follows:
A. 14.7 psi = 1 atm = 760 torr = 760 mm Hg.
B. 1 psi = 0.59 atm = 760 torr = 10 mm Hg.
C. 10 psi = 1 atm = 550 torr = 0.59 mm Hg.
D. 14.7 psi = 100 atm = 30 torr = 100 mm Hg.

______ **10.** The ability of a substance to dissolve in water is known as its:
A. expansion ratio.
B. persistence.
C. water solubility.
D. dispersion value.

______ **11.** pH is an expression of the concentration of:
A. hydrogen ions in a given substance.
B. acid ions in a given substance.
C. oxygen ions in a given substance.
D. base ions in a given substance.

______ **12.** Common acids have pH values that are:
A. equal to zero.
B. greater than 7.
C. equal to 7.
D. less than 7.

______ **13.** Bases have pH values that are:
A. equal to zero.
B. greater than 7.
C. equal to 7.
D. less than 7.

______ **14.** The hazardous chemical compounds released when a material decomposes under heat are known as:
A. carcinogens.
B. alpha particles.
C. toxic products of combustion.
D. beta particles.

______ **15.** The nucleus of a radioactive isotope includes an unstable configuration of:
A. protons and neutrons.
B. electrons and protons.
C. electrons and neutrons.
D. protons, electrons, and neutrons.

______ **16.** Of the following, which is the least penetrating type of radiation?
A. Alpha
B. Beta
C. Gamma
D. Neutron

______ **17.** What is the process by which a person or object transfers contamination to another person or object through direct contact?
A. Contamination by association
B. Secondary exposure
C. Direct contamination
D. Secondary contamination

______ **18.** Exposure to which of the following substances prevents the body from using oxygen?
A. Chlorine
B. Cyanide
C. Lewisite
D. Sarin

_____ **19.** Phosgene is a:
A. nerve agent.
B. blistering agent.
C. choking agent.
D. blood agent.

_____ **20.** Which type of exposure occurs when harmful substances are brought into the body through the respiratory system?
A. Ingestion exposure
B. Inhalation exposure
C. Absorption exposure
D. Injection exposure

_____ **21.** Adverse health effects caused by long-term exposure to a substance are known as:
A. acute health hazards.
B. chronic health hazards.
C. long-term disablers.
D. overexposures.

_____ **22.** Which type of chemical causes a substantial proportion of exposed people to develop an allergic reaction in normal tissue after repeated exposure?
A. Sensitizer
B. Irritant
C. Convulsant
D. Contaminant

Labeling

Label the following diagrams with the correct terms.

1. Vapor density.

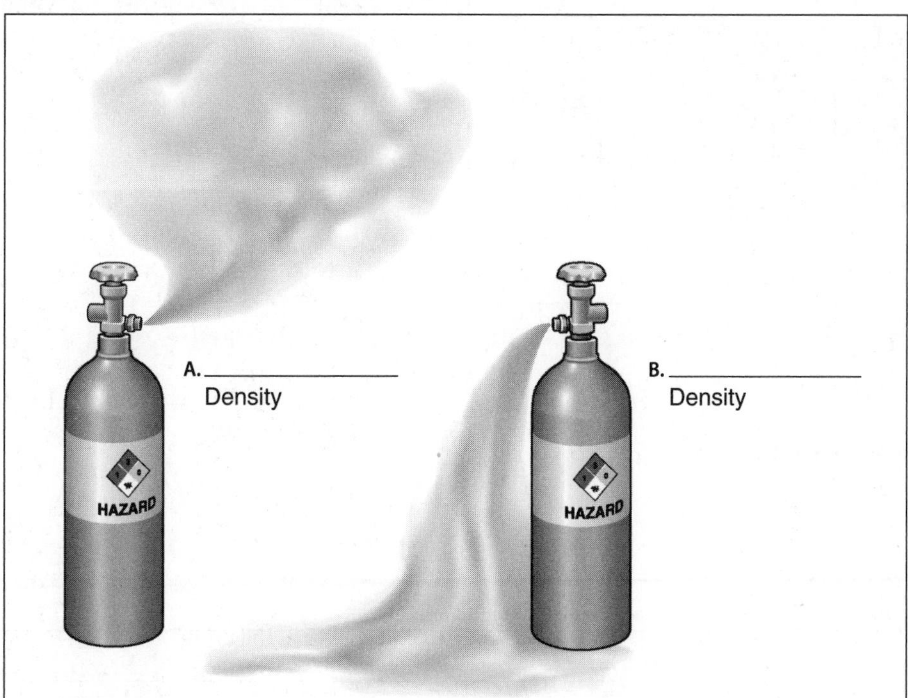

2. Radiation.

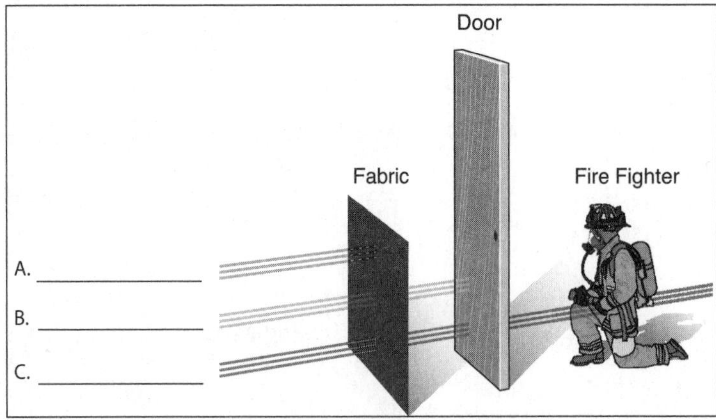

3. Four ways a chemical substance can enter the body.

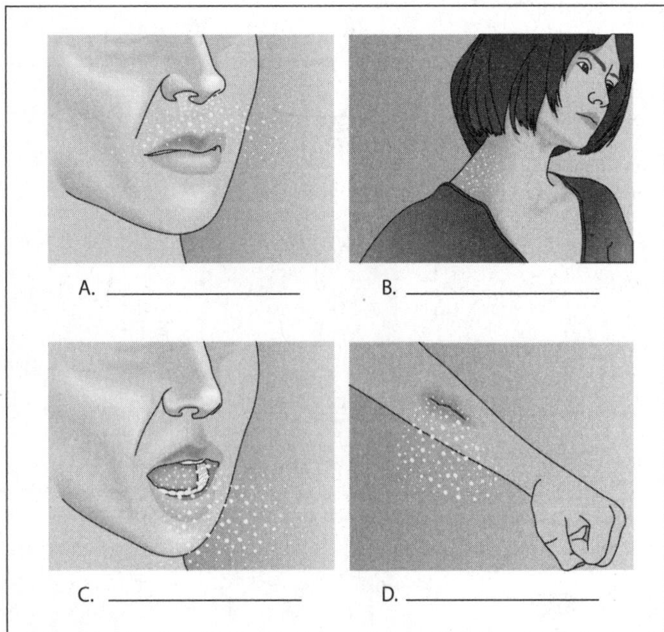

Vocabulary

Define the following terms using the space provided.

1. Absorption:

____

____

____

2. HEPA:

____

____

____

3. Radiation:

____

____

____

4. Contamination:

____

____

____

5. Nerve agent:

____

____

____

Fill-in

Read each item carefully, and then complete the statement by filling in the missing word(s).

1. If the state of matter and physical properties of a chemical are known, a fire fighter can ______________ what the substance will do if it escapes its container.
2. Chemicals can undergo a(n) ______________ change when subjected to outside influences such as heat, cold, and pressure.
3. Standard atmospheric pressure at sea level is ______________ pounds per square inch.
4. The flash point of gasoline is ______________.
5. ______________ ______________ is the minimum temperature at which a liquid or a solid emits vapor sufficient to form an ignitable mixture with air.
6. The ______________ temperature is the minimum temperature at which a substance will ignite without an external ignition source.
7. Most flammable liquids will ______________ on water.
8. The periodic table illustrates all the known ______________ that make up every known compound.
9. ______________ particles can break chemical bonds creating ions; therefore they are considered ionizing radiation.
10. Chemicals that are capable of causing seizures are classified as ______________.

True/False

If you believe the statement to be more true than false, write the letter "T" in the space provided. If you believe the statement to be more false than true, write the letter "F."

1. ______ A physical change is essentially a change in state; a chemical change results in an alteration of the chemical nature of the material.
2. ______ Water has an expansion rate of 100:1 and a boiling point of 100°F (38°C).
3. ______ Diesel fuel has a higher flash point than does gasoline.
4. ______ The wider the flammable range, the more dangerous the material.
5. ______ Vapor pressure directly correlates to the speed at which a material will evaporate once it is released from its container.
6. ______ Radioactive isotopes can be detected by the noise and odors they give off.
7. ______ The nucleus of an atom is made up of protons, neutrons, and electrons.
8. ______ A hazard is a material capable of posing an unreasonable risk to health, safety, or the environment.
9. ______ Nerve agents attack the central nervous system.
10. ______ A chemical brought into the body through an open cut is an injection exposure.

Short Answer

Complete this section with short written answers using the space provided.

1. Identify the seven types of possible hazardous material incidents represented by the mnemonic "TRACEMP."

____

____

____

____

____

____

____

2. Identify the nerve agent signs and symptoms represented by the mnemonic "SLUDGEM."

____

____

____

____

____

____

____

3. Identify and define the four ways through which chemical substances can enter the human body.

4. Identify the two factors that cause radiation to be a health hazard.

5. What does "4H MEDIC ANNA" stand for?

CLUES

Across

7 The least penetrating of the three common types of radiation emitted by radioactive material.
8 A cancer-causing agent.
12 Materials with pH values less than 7.
13 Exposure to a hazardous material by breathing it into the lungs.
14 Exposure to a hazardous material by swallowing it.
15 The process of transferring a hazardous material from its source to people.

Down

1 The gas phase of a substance.
2 A nerve agent that is primarily a vapor hazard.
3 Materials with a pH value greater than 7.
4 High-energy, short-wavelength electromagnetic radiation. (2 words)
5 Hazardous materials entering cuts or other breaches in the skin.
6 The contamination process by which a contaminant is carried out of the hot zone and contaminates additional areas or people.
9 Toxic substances that attack the central nervous system in humans. (2 words)
10 A particle ejected from the nucleus of an unstable atom.
11 Penetrating particles found in the nucleus of the atom that are removed through nuclear fusion or fission.

Word Fun

The following crossword puzzle is an activity provided to reinforce correct spelling and understanding of terminology associated with hazardous materials. Use the clues provided to complete the puzzle.

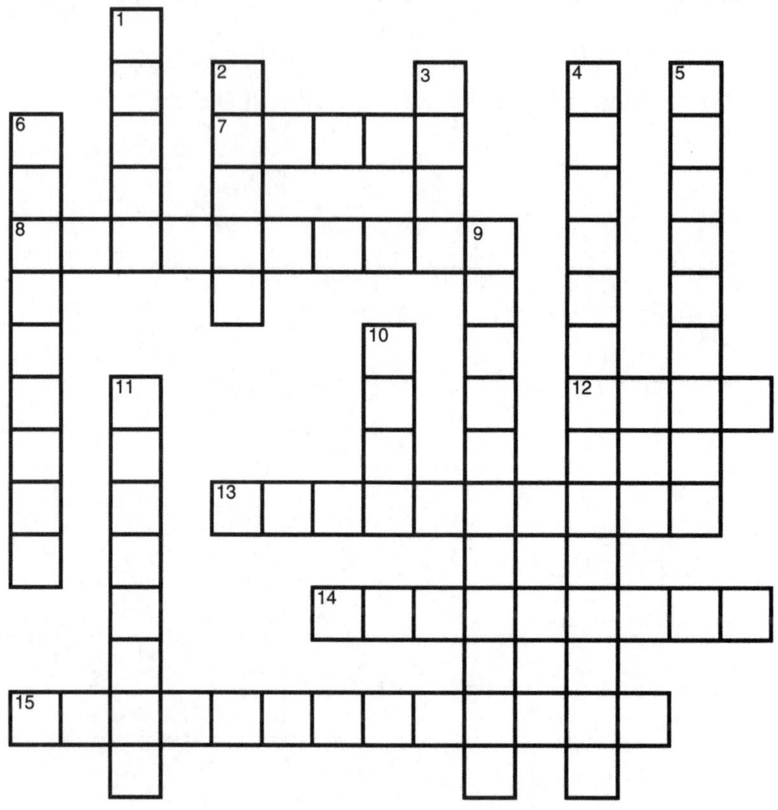

Fire Alarms

The following real case scenarios will give you an opportunity to explore the concerns associated with hazardous materials. Read each scenario, and then answer each question in detail.

1. You are dispatched to the local hospital for a hazardous materials incident. When your engine arrives at the scene, a lab technician states that she believes that one of the medical containers is leaking radioactive material. How should you proceed?

____

____

____

____

____

____

2. It is 2:00 in the afternoon on a Saturday when your engine is dispatched to a shopping mall for a hazardous materials incident. Approximately 50 people have been exposed to a substance that is causing pain and burning to their skin and eyes. Last week, the U.S. terror alert system was activated to level red, and a warning was posted for possible attacks on shopping malls and other highly populated buildings. How should you proceed?

____

____

____

____

____

____

Recognizing and Identifying the Hazards

Workbook Activities

The following activities have been designed to help you. Your instructor may require you to complete some or all of these activities as a regular part of your hazardous materials training program. You are encouraged to complete any activity that your instructor does not assign as a way to enhance your learning in the classroom.

Chapter Review

The following exercises provide an opportunity to refresh your knowledge of this chapter.

Matching

Match each of the terms in the left column to the appropriate definition in the right column.

_____	**1.** Bill of lading	**A.**	The list of every car on a train
_____	**2.** Waybill	**B.**	Barrel-like containers
_____	**3.** Signal words	**C.**	Containers designed to preserve the temperature of the cold liquid held inside
_____	**4.** Dewar containers	**D.**	Portable tanks characterized by a unique style of construction
_____	**5.** Cylinder	**E.**	One or more small openings in closed-head drums
_____	**6.** Consist	**F.**	Shipping papers for trains
_____	**7.** Bungs	**G.**	Information on a pesticide label that indicates the relative toxicity of the material
_____	**8.** Drums	**H.**	Shipping papers for roads and highways
_____	**9.** Vent pipes	**I.**	Inverted J-shaped tubes that allow for pressure relief from the pipeline
_____	**10.** Totes	**J.**	A portable compressed-gas container

Multiple Choice

Read each item carefully, and then select the best response.

_____ **1.** Liquid bulk storage containers have an internal capacity of more than:
- **A.** 500 gallons.
- **B.** 250 gallons.
- **C.** 182 gallons.
- **D.** 119 gallons.

_____ **2.** Solid bulk storage containers have an internal capacity of more than:
- **A.** 1118 pounds.
- **B.** 1084 pounds.
- **C.** 919 pounds.
- **D.** 882 pounds.

_____ **3.** When large-volume horizontal tanks are stored above ground, they are referred to as:
- **A.** OSTs.
- **B.** ASTs.
- **C.** USTs.
- **D.** GSTs.

_____ **4.** Which high-pressure vessels have internal pressures of several hundred pounds per square inch and carry liquefied propane?
A. IM-101 portable tanks
B. IM-102 portable tanks
C. IMO type 10 containers
D. IMO type 5 containers

_____ **5.** Drum bungs can be removed using a:
A. bung wrench.
B. drum ratchet.
C. cinching wrench.
D. drum ring.

_____ **6.** Solids and powders are often stored in:
E. drums.
A. boxes.
B. bags.
C. carboys.

_____ **7.** A glass, plastic, or steel container that holds 5 to 15 gallons of product is a:
A. carboy.
B. bottle.
C. drum.
D. cylinder.

_____ **8.** Propane cylinders contain a liquefied gas and have low pressures of approximately:
A. 50–110 psi.
B. 150–200 psi.
C. 200–300 psi.
D. 300–400 psi.

_____ **9.** Gaseous substances that have been chilled until they liquefy are classified as:
A. cryogenic gases.
B. crystals.
C. Dewar liquids.
D. cryogenic liquids.

_____ **10.** One of the most common chemical tankers is a gasoline tanker, also known as a(n):
A. MC-331 pressure cargo tanker.
B. MC-307 chemical hauler.
C. MC-306 flammable liquid tanker.
D. tube trailer.

_____ **11.** Which types of containers are generally V-shaped with rounded sides and are used to carry grain or fertilizers?
A. Consist tankers
B. Dry bulk cargo tankers
C. Carboys
D. ASTs

_____ **12.** Fire fighters should be able to recognize the three basic railcar configurations of:
A. nonpressurized, pressurized, and special use.
B. dry bulk, liquid, and hazardous materials.
C. agricultural, mechanical, and commercial products.
D. contained, unpackaged, and hazardous materials.

_____ **13.** What are the 10¾" (27-cm) diamond-shaped indicators that must be placed on all four sides of hazardous materials transportation called?
A. MSDS markers
B. Labels
C. Placards
D. NAERG tags

_____ **14.** Within the NFPA hazard identification system, which number is used to identify materials that will not burn?
A. 4
B. 2
C. 1
D. 0

_____ **15.** Within the NFPA hazard identification system, which number is used to identify materials that can cause death after a short exposure?
A. 4
B. 2
C. 1
D. 0

_____ **16.** Shipping papers on a marine vessel are referred to as:
A. waybills.
B. dangerous cargo manifests.
C. consists.
D. freight bills.

_____ **17.** Which DOT packaging group designation is used to represent the highest level of danger?
A. Packaging group I
B. Packaging group II
C. Packaging group III
D. Packaging group V

_____ **18.** If a radiation incident is expected at a fixed facility, which person should be contacted for information?
A. The safety officer
B. The incident commander
C. The facility's radiation safety officer
D. The shift supervisor

_____ **19.** Which type of packaging has an inner containment vessel of glass, plastic, or metal and rubber or vermiculite packaging materials?
A. Type A
B. Type B
C. Type C
D. Type D

_____ **20.** What is the type of clandestine lab most commonly encountered by fire fighters?
A. Paint lab
B. Drug lab
C. Chemical lab
D. Biological lab

Labeling

Label the following diagrams with the correct terms.

1. Chemical transport vehicles.

A. ______________________________

B. ______________________________

C. ______________________________

D. ______________________________

E. ______________________

F. ______________________

G. ______________________

H. ______________________

Vocabulary

Define the following terms using the space provided.

1. Shipping papers:

____
____
____

2. Secondary containment:

____
____
____

3. Hazardous materials:

____
____
____

4. Pipeline right of way:

____
____
____

5. Placards and labels:

____
____
____

Fill-in

Read each item carefully, and then complete the statement by filling in the missing word(s).

1. Scene _______________ is especially important in all hazardous materials incidents.
2. IM- _______________ containers primarily carry flammable liquids and corrosives.
3. MC- _______________ corrosives tankers are used for transporting concentrated nitric acids and other corrosive substances.
4. Compressed gases such as hydrogen, oxygen, and methane are carried by _______________ trailers.
5. Large-diameter _______________ transport natural gas, diesel fuel, and other products from delivery terminals to distribution facilities.
6. Within the NFPA hazard identification system, special hazards that react with water are identified by _______________.
7. The _______________ _______________ _______________ _______________ describes the chemical hazards posed by a particular substance and provides guidance about personal protective equipment employees need to use to protect themselves from workplace hazards.

8. A common source of information about a particular chemical is the ______________ ______________ ______________ ______________ specific to that substance.

9. The ______________ marking system has been developed primarily to identify detonation, fire, and special hazards.

10. Shipping papers for railroad transportation are called ______________; the list of the contents of every car on the train is called a(n) ______________.

True/False

If you believe the statement to be more true than false, write the letter "T" in the space provided. If you believe the statement to be more false than true, write the letter "F."

1. ______ Hazardous materials incidents can occur almost anywhere.
2. ______ Intermodal tanks are both shipping and storage vehicles.
3. ______ Nonbulk storage vessels can also be used as intermodal tanks.
4. ______ Hazardous materials can be transported in cardboard drums or paper bags.
5. ______ The Department of Transportation's marking system is characterized by a system of signs, colors, and numbers.
6. ______ *OX* is used to represent compressed oxygen in the NFPA hazard identification system.
7. ______ *ACID* is used to represent acid in the NFPA hazard identification system.
8. ______ More than 4 billion tons of hazardous materials are shipped annually in the United States.
9. ______ An MSDS will usually include a responsible-party contact.
10. ______ CHEMRESPECT is a free service that connects fire fighters with chemical manufacturers, chemists, and other product specialists who can help during a chemical incident.

Short Answer

Complete this section with short written answers using the space provided.

1. List five pieces of specific information that are included on a pesticide bag label.

____

____

____

____

____

2. Identify the nine *ERG* chemical families.

____

____

____

____

____

____

____

____

____

3. Identify and describe the four colored sections of the *ERG*.

4. Describe the parts and purpose of the NFPA 704 hazard identification system.

5. List five pieces of information that are normally included on a material safety data sheet.

CLUES

Word Fun

The following crossword puzzle is an activity provided to reinforce correct spelling and understanding of terminology associated with hazardous materials. Use the clues provided to complete the puzzle

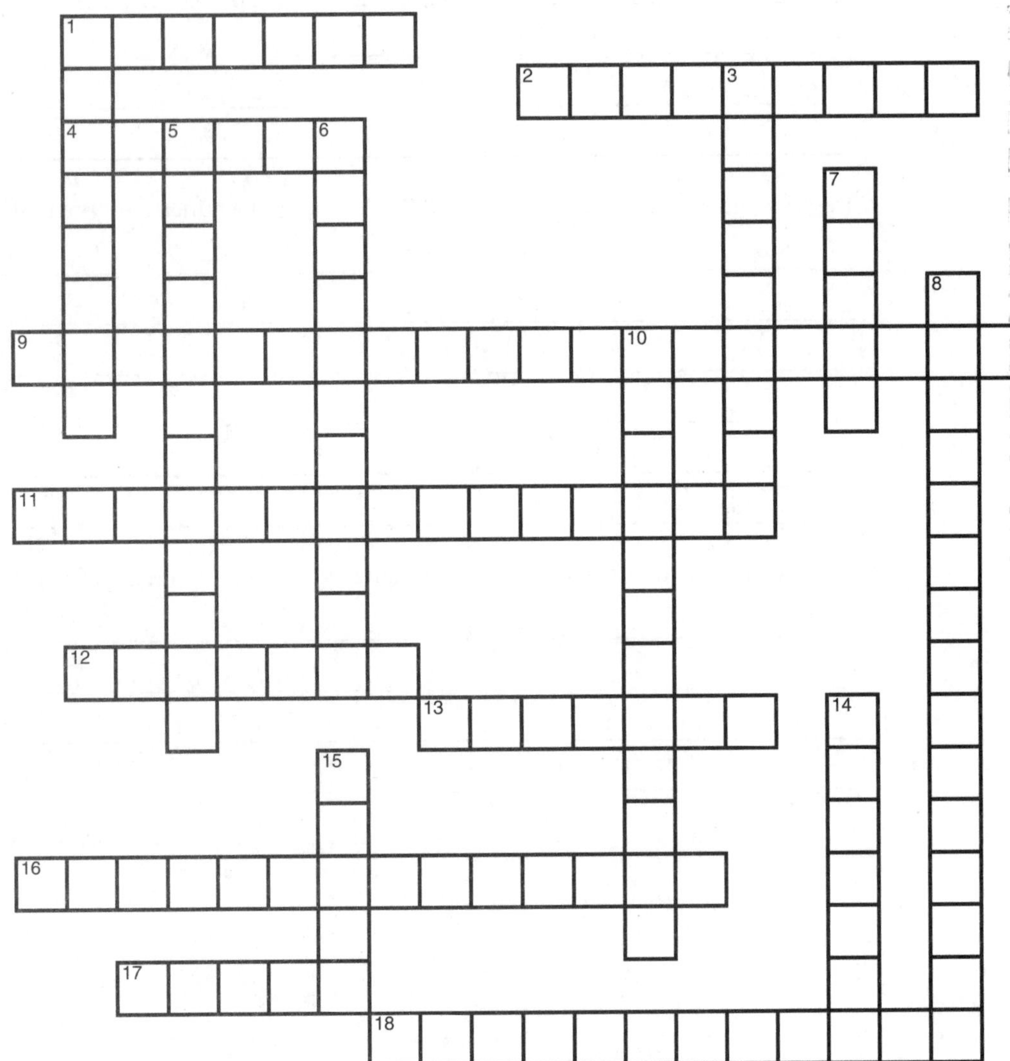

Across

1 A glass, plastic, or steel container, ranging in volume from 5 gallons to 15 gallons.

2 Inverted J-shaped tubes that allow for pressure relief or natural venting of the pipeline for maintenance and repairs. (2 words)

4 Smaller versions of placards, which are placed on four sides of individual boxes and smaller packages.

9 Any device or structure that prevents environmental contamination when the primary container or its appurtenances fail. (2 words)

11 Containers designed to preserve the temperature of the cold liquid held inside them. (2 words)

12 The list of every car on a train.

13 The shipping papers on an airplane. (2 words)

16 A document that usually includes the names and addresses of both the shipper and the receiver, as well as a list of shipped materials along with their quantity and weight. (2 words)

17 Portable tanks, usually holding a few hundred gallons of product, characterized by a unique style of construction.

18 Shipping papers for roads and highways. (2 words)

Down

1 A portable compressed-gas container.

3 A length of pipe, including pumps, valves, flanges, control devices, strainers, and/or similar equipment, for conveying fluids and gases.

5 Shipping papers for roads and highways. (3 words)

6 Information on a pesticide label that indicates the relative toxicity of the material. (2 words)

7 Barrel-like containers built to DOT specification 5P.

8 Bulk containers that can be shipped by all modes of transportation—air, sea, or land. (2 words)

10 High-volume transportation devices made up of several individual compressed-gas cylinders banded together and affixed to a trailer. (2 words)

14 Shipping papers for trains.

15 One or more small openings in closed-head drums.

Fire Alarms

The following real case scenario will give you an opportunity to explore the concerns associated with hazardous materials. Read the scenario, and then answer each question in detail.

1. Your engine company is first on the scene of a vehicle accident on the four-lane highway leading into town. You see an MC-307 lying on its side with liquid leaking from the valving. Your company officer orders you to gather information for a call to CHEMTREC.

   a. What information will you collect for the phone call?

   ____
   ____
   ____
   ____
   ____
   ____

   b. What is the CHEMTREC emergency phone number?

   ____
   ____
   ____
   ____
   ____
   ____

Estimating Potential Harm and Planning a Response

Workbook Activities

The following activities have been designed to help you. Your instructor may require you to complete some or all of these activities as a regular part of your hazardous materials training program. You are encouraged to complete any activity that your instructor does not assign as a way to enhance your learning in the classroom.

Chapter Review

The following exercises provide an opportunity to refresh your knowledge of this chapter.

Matching

Match each of the terms in the left column to the appropriate definition in the right column.

	Term	Definition
_____	**1.** TLV/C	**A.** Examples include diking and damming; stopping the flow of a substance remotely from a valve or shut-off; diluting or diverting material; or suppressing or dispersing vapor
_____	**2.** Defensive objectives	**B.** The maximum concentration of hazardous material to which a worker should not be exposed, even for an instant
_____	**3.** TLV/TWA	**C.** The maximum airborne concentration of a material to which a worker can be exposed for 8 hours a day, 40 hours a week, and not suffer any ill effects
_____	**4.** IDLH	**D.** Describing an atmospheric concentration of a toxic, corrosive, or asphyxiant substance, which poses an immediate threat to life or could cause irreversible or delayed adverse health effects
_____	**5.** REL	**E.** A value established by NIOSH that is comparable to OSHA's PEL

Multiple Choice

Read each item carefully, and then select the best response.

_____ **1.** When planning an initial hazardous materials incident response, what is the first priority?
- **A.** Consider the effect on the environment.
- **B.** Consider the safety of the victims.
- **C.** Consider the equipment and personnel needed to mediate the incident.
- **D.** Consider the safety of the responding personnel.

_____ **2.** Planning a response begins with the:
- **A.** size-up.
- **B.** initial call for help.
- **C.** incident commander's orders.
- **D.** review of standard operating procedures.

_____ **3.** Responders to a hazardous materials incident need to know the:
A. type of material involved.
B. general operating guidelines.
C. short- and long-term effects of the hazardous material.
D. duration of the incident.

_____ **4.** Selection of personal protective equipment is based on the:
A. hazardous material involved.
B. level of training of the responder.
C. direction of the incident commander.
D. standard operating procedures of the department.

_____ **5.** One of the primary objectives of a medical surveillance program is to determine:
A. the intensity of the response at an incident.
B. the concentration of the chemicals at an incident.
C. the time of duty at an incident.
D. any changes in the functioning of body systems.

_____ **6.** When/where do secondary attacks take place?
A. As responders treat victims
B. At the firehouse
C. At the police station
D. Never

_____ **7.** Responders to hazardous materials incidents need to consider:
A. the size of the container.
B. the nature and amount of the material released.
C. the area exposed to the material.
D. A, B, and C are correct.

_____ **8.** Litmus paper is used to determine:
A. the time at which contamination occurred.
B. pH.
C. weather.
D. the location of contamination.

_____ **9.** Victims removed from contaminated zones must be:
A. searched.
B. confined.
C. decontaminated.
D. arrested.

_____ **10.** Defensive actions include:
A. plugging.
B. patching.
C. overpacking.
D. diking.

Labeling

Label the following diagrams with the correct terms.

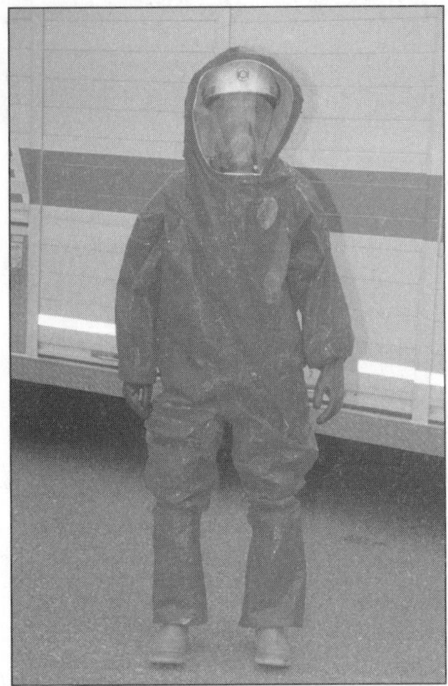

1. Level ______ ensemble

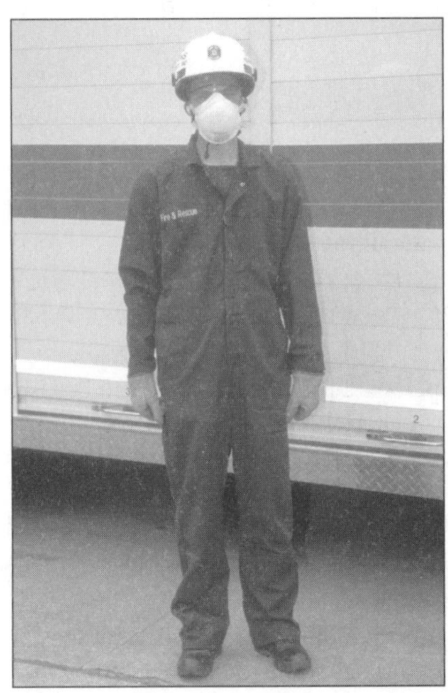

2. Level ______ ensemble

3. Level ______ ensemble

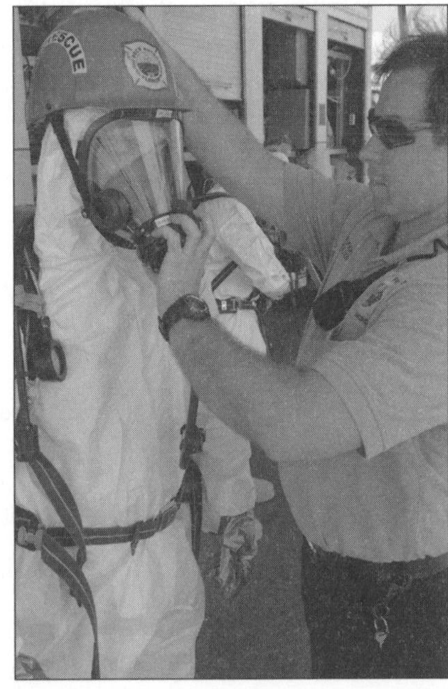

4. Level ______ ensemble

Vocabulary

Define the following terms using the space provided.

1. Defensive objectives:

____

____

____

2. Isolation of the hazard area:

____

____

____

3. Decontamination corridor:

____

____

____

4. Chemical-resistant materials:

____

____

____

5. Supplied-air respirator (SAR):

____

____

____

Fill-in

Read each item carefully, and then complete the statement by filling in the missing word(s).

1. When a hazardous materials incident is detected, there should be an initial call for additional ______________.
2. ______________ burns are often much deeper and more destructive than acid burns.
3. The ______________ of the affected area near the location of the spill or leak are important factors in planning the response to an incident.
4. ______________ ______________ is enhanced by abrasions, cuts, heat, and moisture.
5. The methods of decontamination are dictated by the ______________ ______________.
6. ______________ paper can be used to determine the concentration of an acid or a base by reporting the hazardous material's pH.
7. ______________-_____-______________ is a method of safeguarding people located near or in a hazardous area by temporarily keeping them in a cleaner atmosphere, usually inside structures.

8. Diking and damming, stopping the flow of a substance remotely, and diluting or directing a material are examples of _______________ actions.

9. A primary terrorist attack may purposely injure members of the public to draw _______________ into the scene.

10. _______________ is the process by which a hazardous chemical moves through a given material on the molecular level.

True/False

If you believe the statement to be more true than false, write the letter "T" in the space provided. If you believe the statement to be more false than true, write the letter "F."

1. ______ Degradation and penetration refer to the same process.
2. ______ When dealing with a hazardous material, a variety of sources of information should be compared for consistency.
3. ______ A Level B ensemble provides a high level of respiratory protection but less skin protection than Level A.
4. ______ Level D ensemble is the highest level of protection.
5. ______ Air-purifying respirators (APR) provide breathing air and are appropriate for use when sufficient oxygen for breathing is not available.

Short Answer

Complete this section with short written answers using the space provided.

1. Identify 10 pieces of information that could be reported to agencies to assist in their preparation for a response to a hazardous materials incident.

____

____

____

____

____

____

____

____

____

____

2. Identify the three defensive objectives.

3. List the special technical group that may develop under the Operations section during an incident involving hazardous materials.

4. Identify and define the three basic atmospheres at a hazardous materials emergency according to the exposure guidelines.

5. Identify and provide a brief description of the four levels of protective clothing.

CLUES

Word Fun

The following crossword puzzle is an activity provided to reinforce correct spelling and understanding of terminology associated with hazardous materials. Use the clues provided to complete the puzzle.

Across

4 The atmospheric concentration of any toxic, corrosive, or asphyxiant substance such that it poses an immediate threat to life or could cause irreversible or delayed adverse health effects. (abbreviation)

5 ______________ contamination is the process by which a contaminant is carried out of the hot zone and contaminates people, animals, the environment, or equipment.

6 A policy under which, once the perimeter around a release site has been identified and marked out, responders limit access to all but essential personnel. (3 words)

8 ______________-protective clothing is fully encapsulating chemical protective clothing that offers full-body protection from highly contaminated environments and requires air-supplied respiratory protection devices such as SCBA.

9 A(n) ______________ respirator obtains its air through a hose from a remote source such as a compressor or storage cylinder. (2 words)

10 A respirator with independent air supply used by fire fighters to enter toxic or otherwise dangerous atmospheres. (abbreviation)

11 The removal or relocation of those individuals who may be affected by an approaching release of a hazardous material.

Down

1 The process by which a hazardous chemical moves through a given material on the molecular level.

2 The established standard limit of exposure to a hazardous material. (abbreviation)

3 ______________-protective equipment is a type of personal protective equipment that shields the wearer during short-term exposures to excessive heat. (2 words)

7 ______________ limit value is the point at which a hazardous material or weapon of mass destruction begins to affect a person.

Fire Alarms

The following real case scenarios will give you an opportunity to explore the concerns associated with hazardous materials. Read each scenario, and then answer each question in detail.

1. It is 4:40 in the afternoon when your engine company is dispatched to an industrial plant for a hazardous materials spill. Upon arrival, you see a vapor cloud coming from the rear of the building. The plant has been evacuated. After talking with witnesses who were inside the plant, you learn that the vapor cloud is anhydrous ammonia. A careless forklift operator punctured an unknown-sized storage tank. You are now in the process of gathering the facts and reporting pertinent information to the appropriate agencies. How should you proceed?

2. You are on the scene of a large hazardous material release. Multiple agencies are en route to the site, and it appears that the incident will entail a lengthy response. Recognizing the severity of the incident, the incident commander tells your engine company to secure a location at which to establish a formal command post. How should you proceed?

Implementing the Planned Response

Workbook Activities

The following activities have been designed to help you. Your instructor may require you to complete some or all of these activities as a regular part of your hazardous materials training program. You are encouraged to complete any activity that your instructor does not assign as a way to enhance your learning in the classroom.

Chapter Review

The following exercises provide an opportunity to refresh your knowledge of this chapter.

Matching

Match each of the terms in the left column to the appropriate definition in the right column.

______	**1.** Cold zone	**A.** The decontamination corridor is located here
______	**2.** Backup team	**B.** Gathers information and reports to the incident commander and the safety officer
______	**3.** Public information officer	**C.** The safe area that houses the command post at an incident
______	**4.** Hot zone	**D.** Typically wears the same level of protection as the initial entry team
______	**5.** Unified command	**E.** The point of contact for cooperating agencies on scene
______	**6.** Warm zone	**F.** The point of contact for the media
______	**7.** Liaison officer	**G.** The area immediately surrounding a hazardous materials incident
______	**8.** Technical reference team	**H.** Refers to crews and companies working in the same geographic location
______	**9.** Size-up	**I.** The starting point for implementing any response
______	**10.** Division	**J.** Used where multiple agencies with overlapping jurisdictions or responsibilities are involved in the same incident

Multiple Choice

Read each item carefully, and then select the best response.

______ **1.** Which of the following is one of the first response priorities at a hazardous materials incident?
- **A.** Contacting the property owners
- **B.** Understanding the nature of the problem
- **C.** Containing the hazardous materials
- **D.** Alerting the appropriate responding agencies

______ **2.** To determine how far to extend evacuation distances, hazardous materials technicians should:
- **A.** contact product specialists.
- **B.** refer to the *ERG*.
- **C.** get direction from the incident commander.
- **D.** use detection and monitoring equipment.

CHAPTER

5

______ 3. Which of the following factors is a major concern when considering evacuation?
   A. Potential for exposure to the material
   B. Distance to a safe area
   C. Time of day
   D. Amount of property involved in the incident

______ 4. Which method of safeguarding people in a hazardous area involves keeping them in a safe atmosphere?
   A. Staying indoors
   B. Duck and cover
   C. Shelter-in-place
   D. Containment

______ 5. During the initial size-up at a hazardous materials incident, the first decision concerns:
   A. the amount of property affected.
   B. personnel safety.
   C. the number of people involved.
   D. the type of material involved.

______ 6. Evacuation distances for small spills or fires involving hazardous materials are listed in:
   A. the blue pages of the *ERG*.
   B. the green pages of the *ERG*.
   C. the orange pages of the *ERG*.
   D. the yellow pages of the *ERG*.

______ 7. What is the main hub of the incident management system?
   A. The hot zone
   B. The command post
   C. The staging area
   D. The logistics tent

______ 8. If a person's body temperature falls below 95°F (35°C), he or she may experience:
   A. hypothermia.
   B. death.
   C. hyperthermia.
   D. cold exhaustion.

______ 9. When wet socks are worn at long-term incidents exceeding 12 hours, the wearer may experience:
   A. hypothermia.
   B. trenchfoot.
   C. hyperthermia.
   D. cold exhaustion.

______ 10. After team members undergo decontamination, they should:
   A. prepare for reassignment.
   B. remove all layers of their protective uniforms.
   C. have all vital signs checked.
   D. report to the incident commander.

______ 11. The first step in gaining control of a hazardous materials incident is to isolate the problem and:
   A. equip the cold zone.
   B. keep people away.
   C. establish a backup team.
   D. identify the hazardous materials involved.

_______ **12.** Designated areas at a hazardous materials incident based on safety and the degree of hazard are called:
A. control zones.
B. hot zones.
C. warm zones.
D. cold zones.

_______ **13.** What is the area immediately around and adjacent to the incident?
A. Control zone
B. Hot zone
C. Warm zone
D. Cold zone

_______ **14.** What is the area where personnel and equipment are staged before they enter and after they leave the hot zone?
A. Control zone
B. Hot zone
C. Warm zone
D. Cold zone

_______ **15.** What is the safe area in which personnel do not need to wear any special protective clothing for safe operation?
A. Control zone
B. Hot zone
C. Warm zone
D. Cold zone

Labeling

Label the following diagram with the correct terms.

1. Control zones.

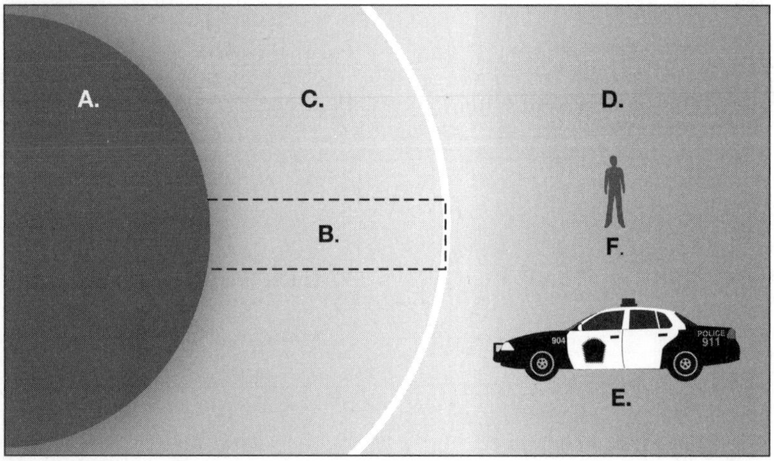

A. ______________________________
B. ______________________________
C. ______________________________
D. ______________________________
E. ______________________________
F. ______________________________

Vocabulary

Define the following terms using the space provided.

1. Shelter-in-place:

____

____

____

2. Heat stroke:

____

____

____

3. Heat exhaustion:

____

____

____

4. Backup personnel:

____

____

____

Fill-in

Read each item carefully, and then complete the statement by filling in the missing word(s).

1. The protection of _______________ is the first priority in any emergency response situation.

2. Before an evacuation order is given, a(n) _______________ _______________ and suitable shelter are established.

3. The _______________ of the hazardous material is a major factor in the decision whether to evacuate.

4. The individual in charge of the ICS shall designate a(n) _______________ _______________ who has a specific responsibility to identify and evaluate hazards.

5. The simplest expression of the _______________ _______________ is for no fewer than two responders to enter a contaminated area.

6. Isolating the scene, protecting the exposures and allowing the incident to stabilize on its own is an example of _______________ actions.

7. Wet clothing extracts heat from the body as many as _______________ times faster than dry clothing.

8. The layer of clothing next to the skin, especially the _______________, should always be kept dry.

9. _______________ is the rapid mental process of evaluating the critical visual indicators of the incident, processing that information, and arriving at a conclusion that formulates your plan of action.

10. Monitoring devices such as wind direction and weather forecasting equipment are critical resources for the _______________ _______________ in formulating response plans.

True/False

If you believe the statement to be more true than false, write the letter "T" in the space provided. If you believe the statement to be more false than true, write the letter "F."

1. ______ The duration of the hazardous materials incident is a factor in determining whether shelter-in-place is a viable option.

2. ______ Monitoring and portable detection devices assist the incident commander in determining the hot, warm, and cold zones and the evacuation distances required.

3. ______ The safety of responders is paramount to maintaining an effective response to any hazardous materials incident.
4. ______ When possible, approach a hazardous materials incident cautiously from downwind of the site.
5. ______ The backup personnel remain on standby in the cold zone awaiting orders to prepare for follow-up duties.
6. ______ The decontamination team must be in place before anyone enters the hot zone.
7. ______ There are several ways to isolate the hazard area and create the control zones.
8. ______ All personnel must be fully briefed before they approach the hazard area or enter the cold zone.
9. ______ The warm zone contains control points for access corridors as well as the decontamination corridor.
10. ______ An incident that involves a gaseous contaminant will require a larger hot zone than one involving a liquid leak.

Short Answer

Complete this section with short written answers using the space provided.

1. Identify and describe some of the key elements of the incident command system.

____

____

____

____

2. Identify and describe the four major functional components within the incident command system.

____

____

____

____

____

3. Identify and provide a brief description of the three zones at a hazardous materials incident.

____

____

____

____

____

Word Fun

The following crossword puzzle is an activity provided to reinforce correct spelling and understanding of terminology associated with hazardous materials. Use the clues provided to complete the puzzle.

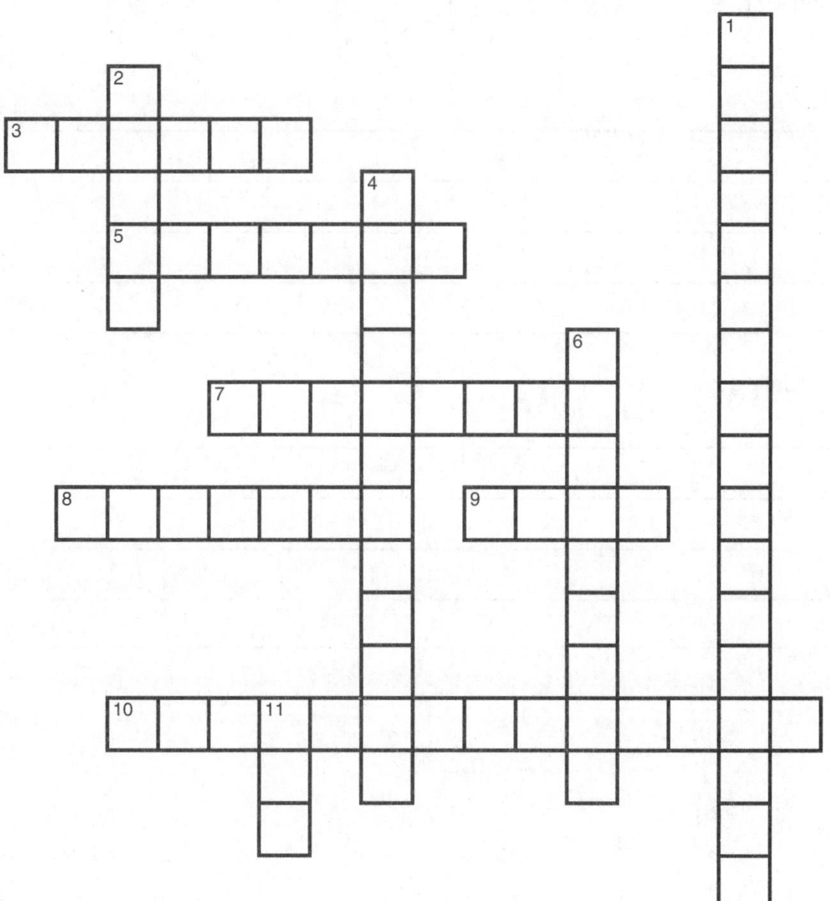

CLUES

Across

3 Heat ________________ is a severe, sometimes fatal condition resulting from the failure of the body's temperature-regulating capacity.

5 ________________ command allows representatives from multiple jurisdictions and agencies to share command authority and responsibility.

7 An organizational level within the incident command system that divides an incident in one location into geographic areas of operational responsibility.

8 A(n) ________________ officer functions as a point of contact with outside agency representatives.

9 The decontamination corridor is located in the ________________ zone.

10 The ________________ officer is responsible for identifying and evaluating hazardous or unsafe conditions at the scene of an incident. (2 words)

Down

1 The ________________ officer functions as a point of contact for the media.

2 A Hazardous Materials ________________ is often established when companies and crews are working on the same task or objective, albeit not necessarily in the same location.

4 The Planning ________________ is responsible for planning functions and for tracking and logging resources.

6 Fully qualified and equipped responders who are assigned to enter into the designated hot zone. (2 words)

11 The location in the cold zone where the command, coordination, control, and communications functions are centralized. (abbreviation)

Fire Alarms

The following real case scenario will give you an opportunity to explore the concerns associated with hazardous materials. Read the scenario, and then answer each question in detail.

1. Your engine company has responded to an explosion at a local abortion clinic. You are instructed to be aware that a secondary device might potentially be present.

a. What is a secondary device?

____

____

____

____

____

b. What are indicators of a potential secondary device?

____

____

____

____

____

Skill Drills

Skill Drill 5-1: Performing an Emergency Decontamination

Test your knowledge of this skill by filling in the correct words in the photo captions.

1. Ensure that you have the appropriate PPE. Stay clear of the product, and do not make physical contact with it. Make an effort to contain ______________. Instruct or assist victims in removing contaminated clothing.

2. Rinse ______________ with copious amounts of water. Avoid using water that is too warm or too cold; room-temperature water is best.

Terrorism

Workbook Activities

The following activities have been designed to help you. Your instructor may require you to complete some or all of these activities as a regular part of your hazardous materials training program. You are encouraged to complete any activity that your instructor does not assign as a way to enhance your learning in the classroom.

Chapter Review

The following exercises provide an opportunity to refresh your knowledge of this chapter.

Matching

Match each of the terms in the left column to the appropriate definition in the right column.

_____	**1.** Phosgene	**A.**	A nerve gas that is both a contact and a vapor hazard and has the odor of camphor
_____	**2.** Secondary device	**B.**	The physical or chemical process of removing any form of contaminant from a person, an object, or the environment
_____	**3.** Soman	**C.**	A yellowish gas that has many industrial uses but also damages the lungs when inhaled
_____	**4.** Decontamination	**D.**	The process of sorting victims based on the severity of injury and medical needs to establish treatment and transportation priorities
_____	**5.** ANFO	**E.**	A chemical agent that causes severe pulmonary damage
_____	**6.** Sulfur mustard	**F.**	A chemical designed to inhibit breathing, which is typically designed to incapacitate rather than kill
_____	**7.** Cyanide	**G.**	Toxic substances that attack the central nervous system in humans
_____	**8.** Radiological agents	**H.**	Time period between the initial infection by an organism and the development of symptoms
_____	**9.** Nerve agent	**I.**	A weapon intended to cause mass casualties, damage, and chaos
_____	**10.** Choking agent	**J.**	Materials that emit radioactivity
_____	**11.** Chlorine	**K.**	An explosive material containing ammonium nitrate fertilizer and fuel oil
_____	**12.** Sarin	**L.**	A nerve agent that is primarily a vapor hazard
_____	**13.** Triage	**M.**	An explosive device designed to injure emergency responders who have responded to an initial event
_____	**14.** Incubation period	**N.**	A clear, yellow, or amber oily liquid with a faint sweet odor of mustard or garlic that may be dispersed in an aerosol form
_____	**15.** Weapons of mass destruction (WMD)	**O.**	A highly toxic chemical agent that attacks the circulatory system

CHAPTER

6

Multiple Choice

Read each item carefully, and then select the best response.

_____ **1.** A terrorist threat requires fire fighters to work closely with:
A. local, state, and federal law enforcement agencies.
B. emergency management agencies.
C. the military.
D. All of the above.

_____ **2.** Bombing a store that sells fur coats would be an example of:
A. ecoterrorism.
B. cyberterrorism.
C. agroterrorism.
D. religious terrorism.

_____ **3.** Disrupting or deleting government or banking computer systems is an example of:
A. ecoterrorism.
B. cyberterrorism.
C. agroterrorism.
D. religious terrorism.

_____ **4.** Attacking a food industry or supply is an example of:
A. ecoterrorism.
B. cyberterrorism.
C. agroterrorism.
D. religious terrorism.

_____ **5.** An IED is an explosive device that is contained in a package. IED is an acronym for:
A. improvised explosive device.
B. internal explosive device.
C. imploding explosive device.
D. illuminating explosive device.

_____ **6.** At an incident where there is potential terrorist or secondary device activity, the fire department should be part of a joint command structure commonly referred to as:
A. a team command.
B. an emergency response team.
C. a unified command.
D. a united command.

_____ **7.** During a bomb disposal, where does the rapid intervention team stand by to provide immediate assistance?
A. Bomb disposal containment area
B. Forward staging area
C. Incident command center
D. Response area

_____ **8.** Before anyone is allowed to enter a building involved in an explosion, what must happen?
A. The utilities must be disconnected.
B. All emergency response teams must arrive.
C. Team members must review preincident plans.
D. The stability of the building must be evaluated.

______ 9. Many of the chemicals classified as weapons of mass destruction (WMD) are:
  A. expensive.
  B. easy to obtain.
  C. restricted under the Anti-terrorist Act.
  D. kept away from ordinary citizen contact.

______ 10. Which name is given to the time period between the actual infection and the appearance of symptoms?
  A. Growth period
  B. Dispersing period
  C. Incubation period
  D. Implementation period

______ 11. What are the three types of radiation?
  A. Internal, external, and dispersion
  B. Alpha particles, beta particles, and gamma rays
  C. Alpha, beta, and gamma particles
  D. Alpha particles, beta particles, and sigma rays

______ 12. For what purpose is a personal dosimeter used?
  A. To record personal exposure to contaminants
  B. To document personal exposure to contaminants
  C. To measure the amount of radioactive exposure
  D. To measure the active agents in the area

______ 13. Fire fighters responding to a potential or known terrorist incident should use the same approach as they would use for a(n):
  A. structural fire.
  B. EMS incident.
  C. rescue incident.
  D. hazardous materials incident.

______ 14. Fire fighters and emergency responders must remember that a terrorist incident is also a(n):
  A. crime scene.
  B. opportunity to improve working relations between departments.
  C. implementation of advanced rescue techniques.
  D. All of the above.

______ 15. What is the process of sorting victims based on the severity of their injuries and medical needs to establish treatment and transportation priorities called?
  A. EMS
  B. Decon
  C. Triage
  D. Beta

Labeling

Label the following diagrams with the correct terms.

1. Based on the symptoms shown in these photographs, what type of agent was each victim exposed to?

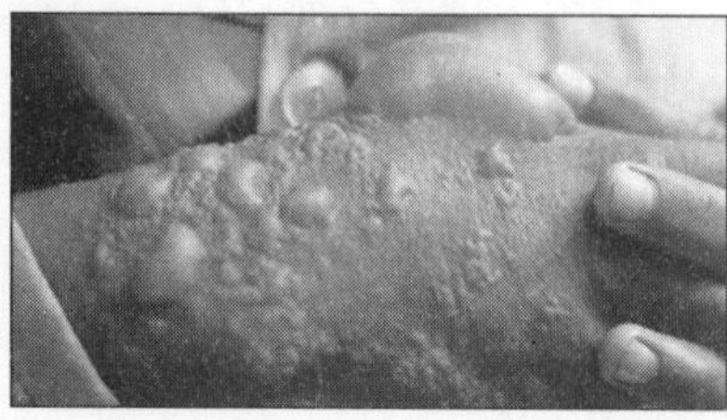

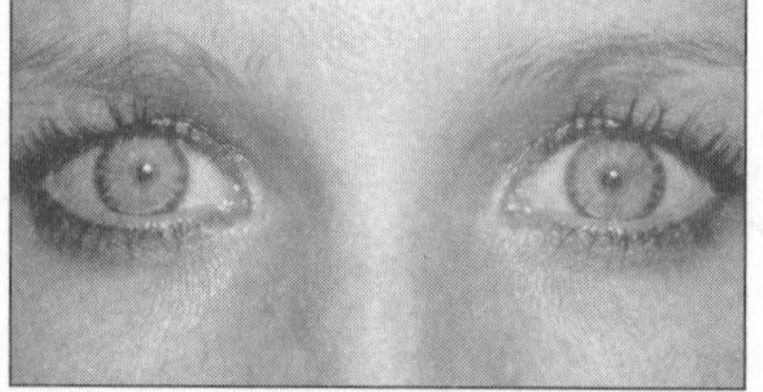

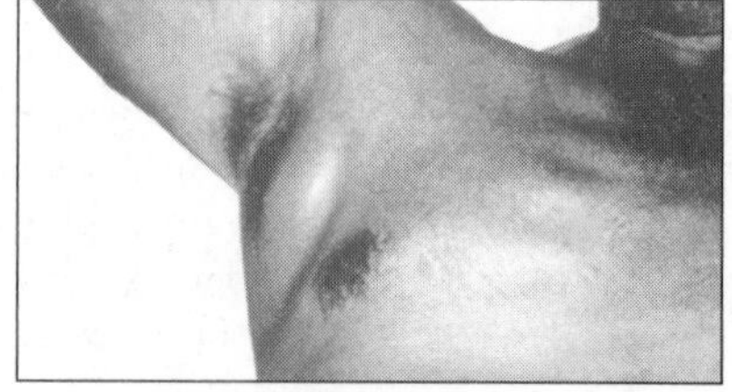

A. ______________________

B. ______________________

C. ______________________

Vocabulary

Define the following terms using the space provided.

1. V-agent:

2. Plague:

3. Smallpox:

4. Tabun:

5. Universal precautions:

6. Forward staging area:

7. Radiation dispersal device:

Fill-in

Read each item carefully, and then complete the statement by filling in the missing word(s).

1. _______________ can be described as the unlawful use of violence or threats of violence to intimidate or coerce a person or group to further political or social objectives.
2. The most common improvised explosive device is the _______________.
3. _______________ _______________ are toxic substances used to attack the central nervous system and were first developed in Germany before World War II.
4. _______________ is a mnemonic used to remember the symptoms of possible nerve agent exposure.
5. _______________-_______________ equipment could be used to distribute chemical agents.
6. The time period between the actual infection and the appearance of symptoms is known as the _______________ _______________.
7. _______________ _______________ release energy in the form of electromagnetic waves or energy particles that cannot be detected by the senses.
8. Decontamination should occur as soon as possible to prevent further _______________ of a contaminant and to reduce the possibility of spreading the contamination.
9. Decontamination of a large number of victims or emergency responders is referred to as _______________ _______________.
10. If contamination is suspected, a plan must ensure that it does not spread beyond a defined _______________.

True/False

If you believe the statement to be more true than false, write the letter "T" in the space provided. If you believe the statement to be more false than true, write the letter "F."

1. _______ Anthrax and the plague are examples of nerve agents.
2. _______ Emergency responders are decontaminated once they leave the contaminated area.
3. _______ Gamma rays are the least harmful of the three types of radiation.
4. _______ Beta particles are also active nerve agents.
5. _______ A personal dosimeter is used to measure the amount of radioactive exposure.
6. _______ Soman is a highly infectious disease that kills approximately 30 percent of those infected with it.
7. _______ Universal precautions must be taken when responding to acts of cyberterrorism.
8. _______ Fire fighters must become familiar with potential terrorist targets and actions because they are often involved in the initial response and handling of a terrorist incident.
9. _______ The first emergency response units to arrive should establish an outer perimeter to control access to and from the scene.
10. _______ Exposure to high levels of radiation can cause vomiting and digestive system damage within a short time.

Short Answer

Complete this section with short written answers using the space provided.

1. What motivates terrorists?

2. Describe ecoterrorism, cyberterrorism, and agroterrorism.

3. Describe the issues that fire fighters must consider following a large explosion.

4. Describe why responding to a terrorist incident puts fire fighters and emergency personnel at increased risk.

5. Identify the three ways to limit exposure to radioactivity.

CLUES

Word Fun

The following crossword puzzle is an activity provided to reinforce correct spelling and understanding of terminology associated with hazardous materials. Use the clues provided to complete the puzzle.

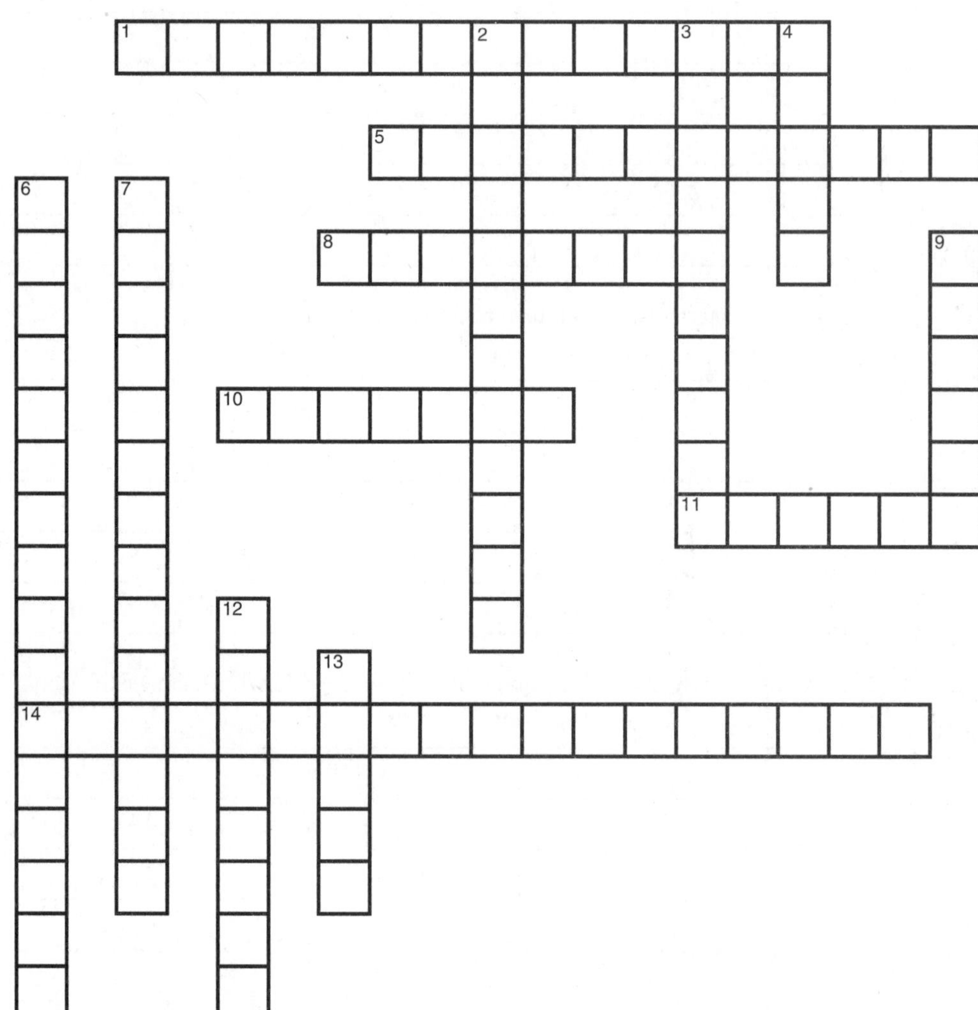

Across

1 Disease-causing bacteria, viruses, and other agents that attack the human body. (2 words)

5 Terrorism directed against causes that radical environmentalists think would damage the earth or its creatures.

8 A blister-forming agent that is an oily, colorless to dark-brown liquid with an odor of geraniums.

10 An infectious disease spread by the bacteria *Bacillus anthracis*; it is typically found around farms, infecting livestock.

11 The process of sorting victims based on the severity of injury and medical needs to establish treatment and transportation priorities.

14 Devices that measure the amount of radioactive exposure to an individual. (2 words)

Down

2 A chemical designed to inhibit breathing and typically intended to incapacitate rather than kill. (2 words)

3 A toxic substance that attacks the central nervous system in humans. (2 words)

4 A nerve agent that is primarily a vapor hazard.

6 The time period between the initial infection by an organism and the development of symptoms. (2 words)

7 The intentional act of electronically attacking government or private computer systems.

9 An infectious disease caused by the bacterium *Yersinia pestis*; it is commonly found on rodents.

12 A chemical agent that causes severe pulmonary damage.

13 A type of radiation that can travel significant distances, penetrating most materials and passing through the body.

Fire Alarms

The following real case scenarios will give you an opportunity to explore the concerns associated with hazardous materials. Read each scenario, and then answer each question in detail.

1. Your fire department contracts with a small farming community for fire protection and EMS. The government has just announced in a press release that it has uncovered some information that agroterrorists are planning a chemical attack on the U.S. food supply. During your shift meeting, your Lieutenant reviewed the department's response guidelines to terrorist incidents. What is the fire department's role in protecting communities from terrorism?

____

____

____

____

____

____

2. It is 11:30 on a Saturday morning when your engine is dispatched to a car explosion in front of a government building near the center of your community. Upon arrival, you find a heavily damaged passenger vehicle that is on fire. The building's widows have been blown out in the front and the structural status is unknown. You observe approximately 10 people who are injured. How should you proceed?

____

____

____

____

____

____

Mission-Specific Competencies: Personal Protective Equipment

Workbook Activities

The following activities have been designed to help you. Your instructor may require you to complete some or all of these activities as a regular part of your hazardous materials training program. You are encouraged to complete any activity that your instructor does not assign as a way to enhance your learning in the classroom.

Chapter Review

The following exercises provide an opportunity to refresh your knowledge of this chapter.

Matching

Match each of the terms in the left column to the appropriate definition in the right column.

_____	**1.** HEPA	**A.**	Offers the lowest level of protection
_____	**2.** Degradation	**B.**	Filters that catch particles down to 0.3-micron size
_____	**3.** Dehydration	**C.**	The flow or movement of a hazardous chemical through closures such as zippers, seams, or imperfections in material
_____	**4.** Level D ensemble	**D.**	Usually precedes heat exhaustion, heat cramps, and heat stroke
_____	**5.** Permeation	**E.**	The physical destruction of clothing material as a result of chemical exposure
_____	**6.** Penetration	**F.**	Devices worn to filter particulates and contaminants from the air
_____	**7.** APRs	**G.**	A process that is similar to water saturating a sponge
_____	**8.** SARs	**H.**	Worn when the airborne substance is known, criteria for APR are met, and skin and eye exposure is unlikely
_____	**9.** Level A ensemble	**I.**	Useful during extended operations such as decontamination, clean-up, and remedial work
_____	**10.** Level C ensemble	**J.**	Used when the hazardous material identified requires the highest level of protection for skin, eyes, and respiration

Multiple Choice

Read each item carefully, and then select the best response.

_____ **1.** Time, distance, and shielding are the preferred methods of protection for which of the following?
- **A.** Biological exposures
- **B.** Chemical exposures
- **C.** Radiation exposures
- **D.** High thermal exposures

_____ **2.** The process by which a hazardous chemical moves through closures, seams, or porous materials is called:
- **A.** penetration.
- **B.** degradation.
- **C.** permeation.
- **D.** vaporization.

______ **3.** The physical destruction of clothing due to chemical exposure is called:
A. penetration.
B. degradation.
C. permeation.
D. vaporization.

______ **4.** Chemical resistance, flexibility, abrasion, temperature resistance, shelf life, and sizing criteria are requirements that need to be considered when selecting:
A. entry tools.
B. respirators.
C. testing equipment.
D. chemical-protective clothing.

______ **5.** Air-purifying respirators should be worn in atmospheres where the type and quantity of contaminants are:
A. unknown.
B. known.
C. suspected.
D. indistinguishable.

______ **6.** Chemical-protective clothing is rated for its effectiveness against chemical permeation, including how quickly it protects the fire fighter and:
A. how well it fits the fire fighter.
B. how many times the suit can be used.
C. to what degree it protects the fire fighter.
D. how visible the fire fighter is when wearing the suit.

______ **7.** Vapor-protective clothing requires the wearer to use:
A. air-purifying respirators.
B. powered air-purifying respirators.
C. particulate respirators.
D. self-contained breathing apparatus.

______ **8.** All of the following may impact the service life of chemical-resistant materials, except:
A. permeation.
B. degradation.
C. penetration.
D. sublimation.

______ **9.** A response to an ammonia leak inside a poorly ventilated storage area within an ice-making facility might require responders to wear:
A. vapor-protective clothing.
B. liquid splash–protective clothing.
C. high temperature–protective equipment.
D. structural fire protective clothing.

______ **10.** The *Standard on Vapor-Protective Ensembles for Hazardous Materials Emergencies* has been established by:
A. OSHA.
B. NFPA.
C. EPA.
D. NIOSH.

Labeling

Label the following diagrams with the correct terms.

1. Types of personal protective equipment.

A. ______________________ **B.** ______________________ **C.** ______________________

Vocabulary

Define the following terms using the space provided.

1. Level A ensemble:

____

____

____

2. Allied professional:

____

____

____

3. Dehydration:

____

____

____

4. Donning:

____

____

____

5. High temperature–protective equipment:

____

____

____

Fill-in

Read each item carefully, and then complete the statement by filling in the missing word(s).

1. Work uniforms offer the ______________ amount of protection in a hazardous materials emergency.
2. ______________ are most likely to penetrate material.
3. ______________-protective clothing is designed to prevent chemicals from coming in contact with the body and may have varying degrees of resistance.
4. An encapsulated suit is a(n) ______________ piece garment that completely encloses the wearer.
5. Forced-air cooling systems, ice-cooled or gel-packed vests, and fluid-chilled systems are examples of ______________-______________ units.
6. ______________ radiation is so strong that structural fire fighters' protective gear and/or any chemical-protective clothing will offer no significant level of protection from this threat.
7. ______________ is an acronym used to sum up a collection of potential hazards that an emergency responder may face.
8. ______________ particles have weight and mass and cannot travel very far from the nucleus of the atom.
9. Fire fighters should be encouraged to drink ______________ of water before donning any protective clothing.
10. Levels A, B, C, and D ensemble classifications have been established by EPA and OSHA ______________ regulations.

True/False

If you believe the statement to be more true than false, write the letter "T" in the space provided. If you believe the statement to be more false than true, write the letter "F."

1. ______ Beta particles are more energetic than alpha particles and pose a greater health hazard.
2. ______ Standard firefighting turnout gear offers little chemical protection, but does have a high degree of abrasion resistance and prevents direct skin contact.
3. ______ Tyvek provides satisfactory protection from all chemicals.
4. ______ Vapor-protective clothing and chemical-protective clothing are identical.
5. ______ Degradation is the process by which a chemical moves through a given material on the molecular level.
6. ______ Reusable Level A suits are required to be pressure-tested after each use.
7. ______ Limited use chemical-protective clothing is expected to be discarded along with the other hazardous waste generated by the incident.
8. ______ HEPA filters are effective for anthrax spores.
9. ______ Powered air-purifying respirators are effective for protection against asphyxiants such as nitrogen and helium.
10. ______ The rapid and destructive way that gasoline dissolves a Styrofoam cup is an example of degradation.

Short Answer

Complete this section with short written answers using the space provided.

1. What seven hazards does the acronym TRACEMP refer to?

2. Identify and provide a brief description of the four levels of protective clothing.

3. What types of allied professionals might assist hazardous materials emergency responders?

Word Fun

The following crossword puzzle is an activity provided to reinforce correct spelling and understanding of terminology associated with hazardous materials. Use the clues provided to complete the puzzle.

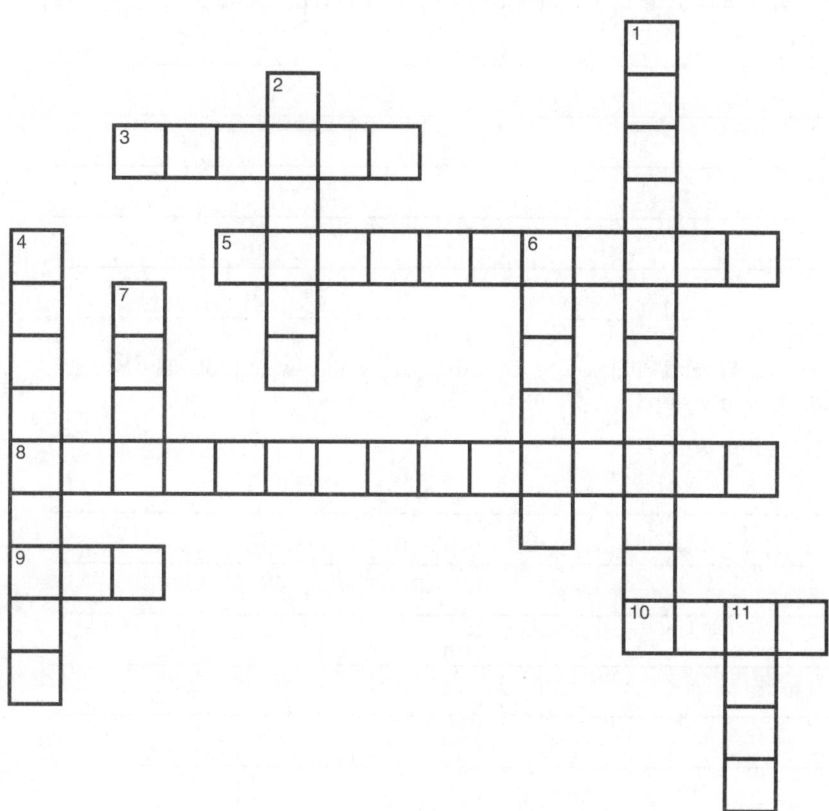

CLUES

Across

3 A(n) ______________ ensemble is used when the atmosphere contains no known hazard, and work functions preclude splashes, immersion, or the potential for unexpected inhalation of or contact with hazardous levels of chemicals; primarily a work uniform that includes coveralls and affords minimal protection.

5 An excessive loss of body water.

8 The National Institute for Occupational ______________ sets the design, testing, and certification requirements for self-contained breathing apparatus in the United States. (3 words)

9 A device worn to filter particulates and contaminants from the air before it is inhaled. (abbreviation)

10 A(n) ______________ filter is used in conjunction with SCBA or simple respirators; catches particles down to 0.3-micron size—much smaller than a typical dust particle or anthrax spore. (abbreviation)

Down

1 ______________-protective clothing is designed to protect the wearer from chemical splashes. (2 words)

2 A(n) ______________ ensemble is used when the type and atmospheric concentration of substances requires a high level of respiratory protection but less skin protection.

4 Chemical ______________ materials are specifically designed to inhibit the passage of chemicals into and through the material by the process of penetration, permeation, or degradation.

6 A person with unique skills, knowledge, and/or abilities, who may be called upon to assist hazardous materials responders, is considered to be a(n) ______________ professional.

7 To ______________ describes the act of taking off an ensemble of PPE.

11 A type of air-purifying respirator that uses a battery-powered blower to pass outside air through a filter and then to the mask via a low-pressure hose.

Fire Alarms

The following real case scenarios will give you an opportunity to explore the concerns associated with hazardous materials. Read each scenario, and then answer each question in detail.

1. During hazardous materials response training, you are assigned to an entry team wearing Level B nonencapsulated personal protective clothing. What is the recommended PPE for Level B protection?

2. After 45 minutes of training while wearing your Level B PPE, you begin feeling dizzy and sweat profusely. You are also feeling weak and notice some blurring of your vision.

   a. What is the probable cause of your sudden illness?

   b. Which actions should be taken?

Skill Drills

Skill Drill 7-1: Donning a Level A Ensemble

Test your knowledge of this skill drill by placing the photos below in the correct order. Number the first step with a "1," the second step with a "2," and so on.

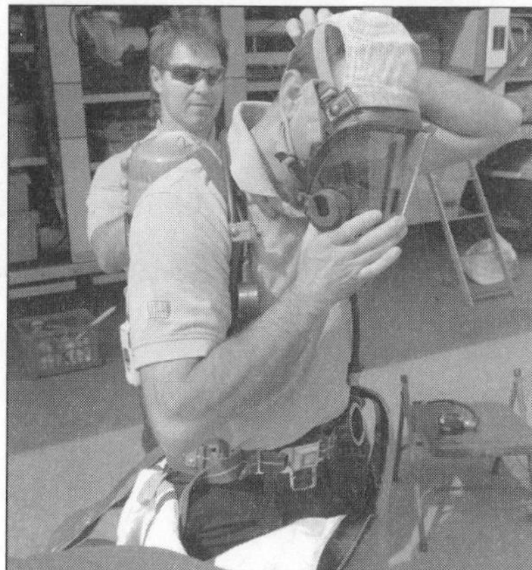

______ Stand up and don the SCBA frame and SCBA face piece, but do not connect the regulator to the face piece.

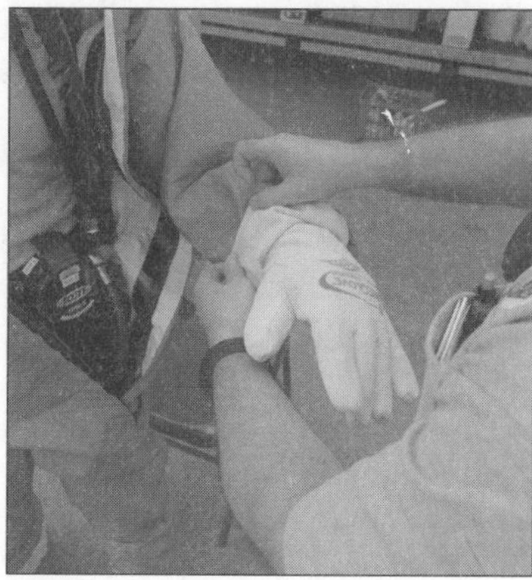

______ Don the outer chemical gloves (if required by the manufacturer's specifications). With assistance, complete donning the suit by placing both arms in the suit, pulling the expanded back piece over the SCBA, and placing the chemical suit over the head.

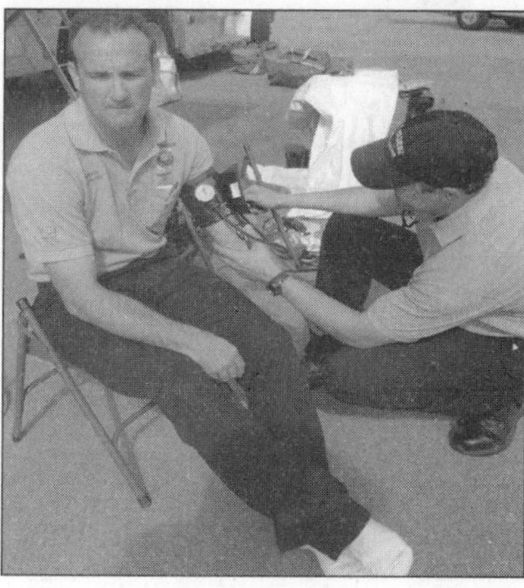

______ Conduct a pre-entry briefing, medical monitoring, and equipment inspection.

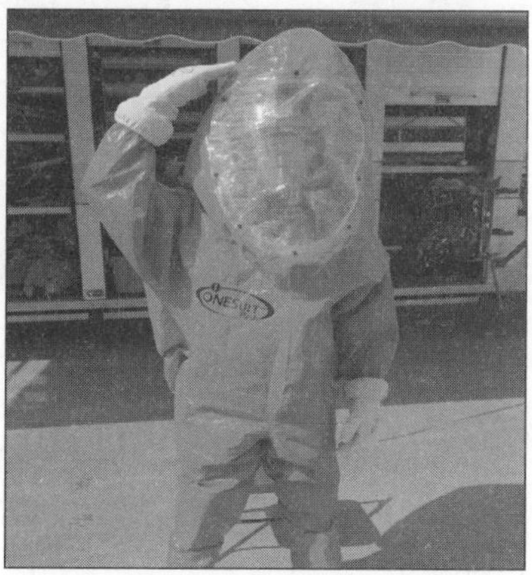

______ Review hand signals and indicate that you are okay.

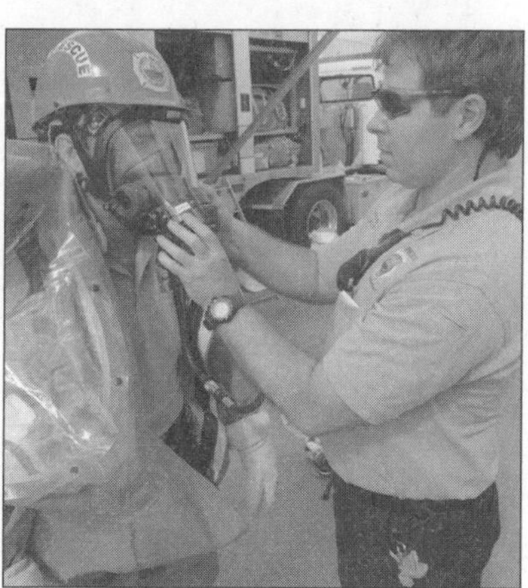

______ Instruct the assistant to connect the regulator to the SCBA face piece and ensure air flow.

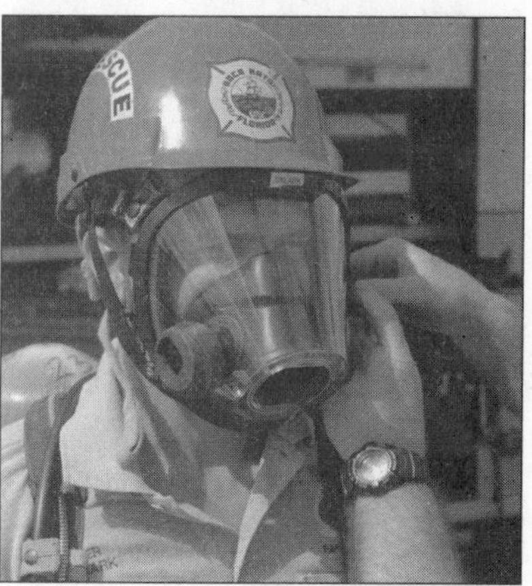

______ Place the helmet on the head.

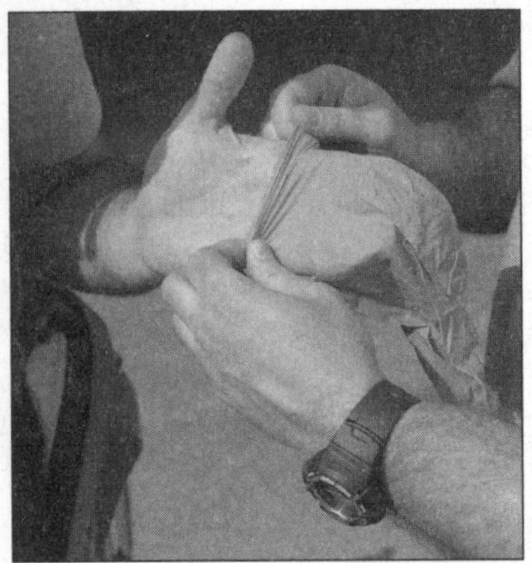

______ Don the inner gloves.

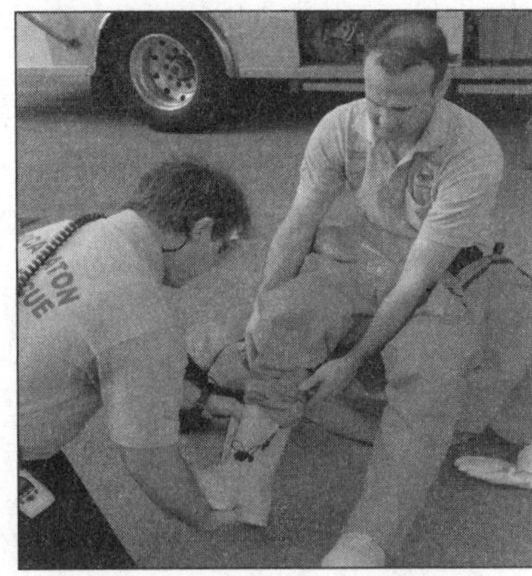

______ While seated, pull on the suit to waist level; pull on the chemical boots over the top of the chemical suit. Pull the suit boot covers over the tops of the boots.

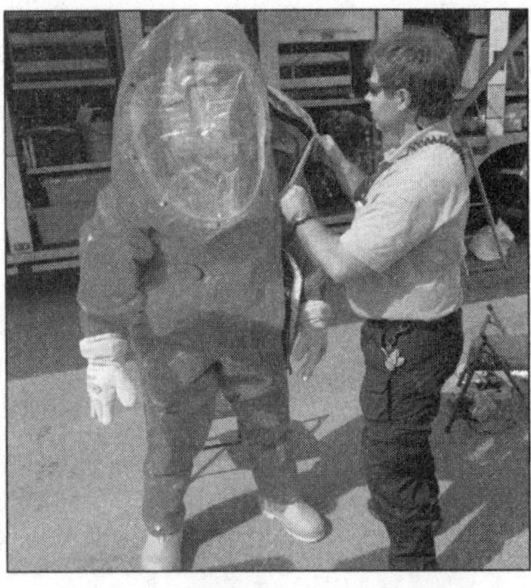

______ Instruct the assistant to close the chemical suit by closing the zipper and sealing the splash flap.

Skill Drill 7-2: Doffing a Level A Ensemble

Test your knowledge of this skill drill by filling in the correct words in the photo captions.

1. After completing decontamination, proceed to the clean area for suit doffing. Pull the hands out of the ______________ ______________ and arms from the sleeves, and cross the arms in front inside the suit.

2. Instruct the assistant to open the ______________ ______________ flap and suit zipper.

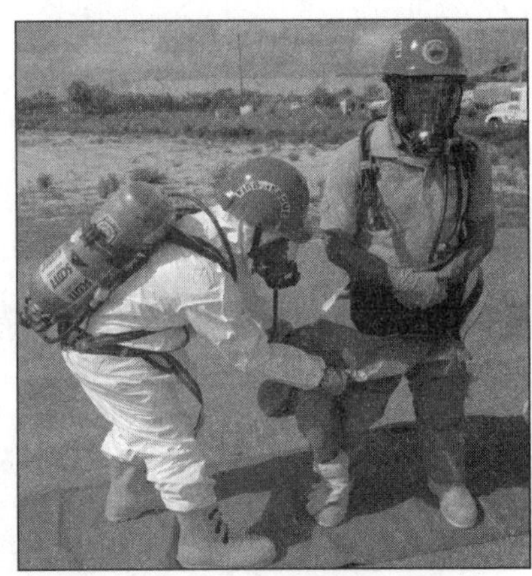

3. Instruct the assistant to begin at the head and roll the suit ______________ and ______________ until the suit is below waist level.

4. Instruct the assistant to complete rolling the suit from the waist to the ankles; step out of the attached _______________ _______________ and suit.

5. Doff the SCBA frame. The _______________ _______________ should be kept in place while the SCBA frame is doffed.

6. Take a deep breath and doff the SCBA face piece; carefully peel off the inner gloves, and walk away from the clean area. Go to the rehabilitation area for _______________ _______________, rehydration, and personal decontamination shower.

Skill Drill 7-3: Donning a Level B Nonencapsulated Chemical-Protective Clothing Ensemble

Test your knowledge of this skill drill by placing the photos below in the correct order. Number the first step with a "1," the second step with a "2," and so on.

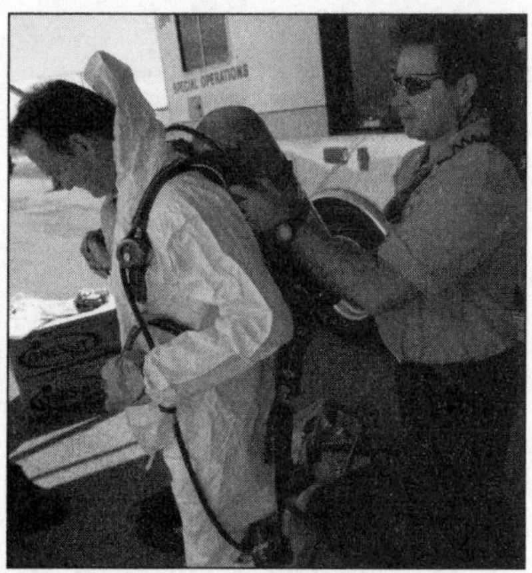

______ Don the SCBA frame and SCBA face piece, but do not connect the regulator to the face piece.

______ While seated, pull on the suit to waist level; pull on the chemical boots over the top of the chemical suit. Pull the suit boot covers over the tops of the boots.

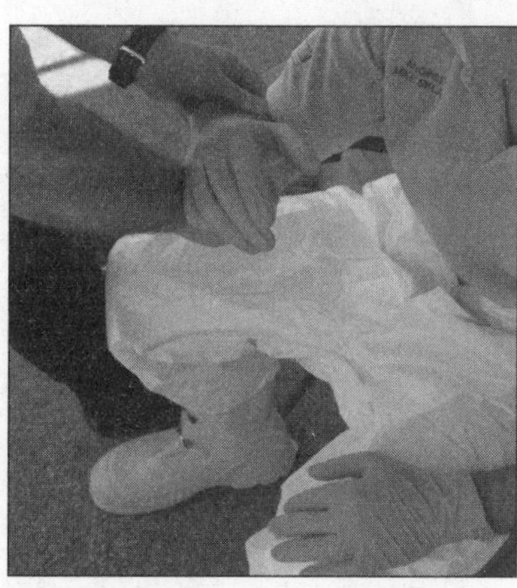

______ Don the inner gloves.

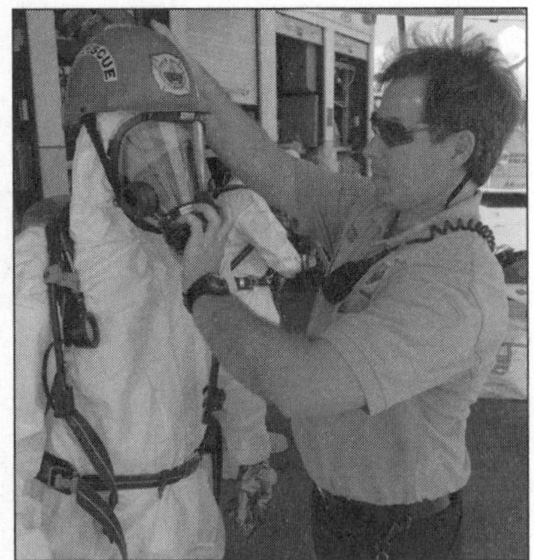

______ With assistance, pull the hood over the head and SCBA face piece. Place the helmet on the head. Put on the outer gloves. Instruct the assistant to connect the regulator to the SCBA face piece and ensure you have air flow.

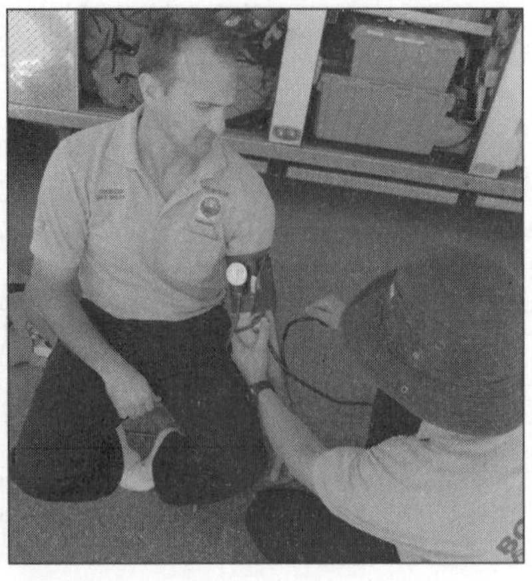

______ Conduct a pre-entry briefing, medical monitoring, and equipment inspection.

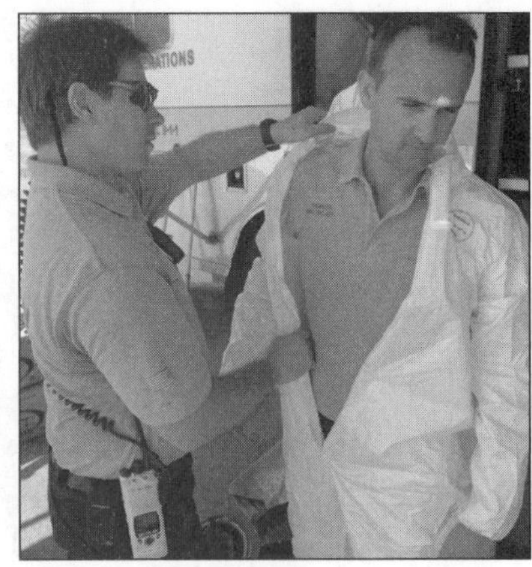

______ With assistance, complete donning the suit by placing both arms in the suit and pulling the suit over the shoulders. Instruct the assistant to close the chemical suit by closing the zipper and sealing the splash flap.

Skill Drill 7-4: Doffing a Level B Nonencapsulated Chemical-Protective Clothing Ensemble

Test your knowledge of this skill drill by filling in the correct words in the photo captions.

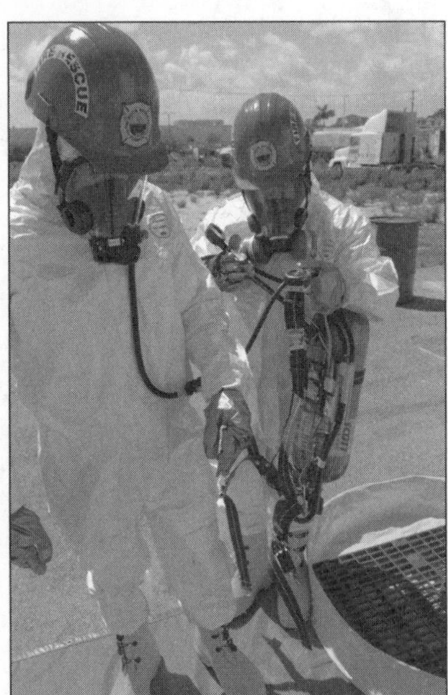

1. After completing decontamination, proceed to the ______________ ______________ for suit doffing. Stand and doff the SCBA frame. Keep the face piece in place.

2. Instruct the assistant to open the chemical splash flap and open the ______________ ______________.

3. Remove your hands from the outer gloves and arms from the sleeves of the suit. Cross your arms in front ______________ the suit. Instruct the assistant to begin at the head and roll the suit down and away until the suit is below waist level.

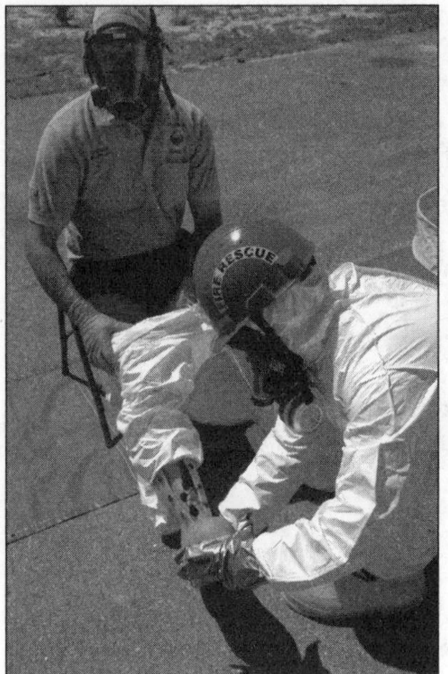

4. Sit down and instruct the assistant to complete rolling down the suit to the _______________; step out of the attached chemical boots and suit.

5. Stand and doff the SCBA face piece and _______________.

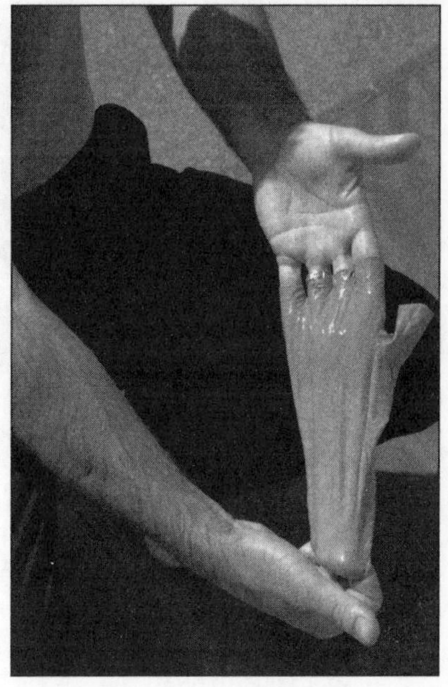

6. Carefully peel off the inner gloves, and walk away from the clean area. Go to the rehabilitation area for medical monitoring, rehydration, and _______________ _______________ _______________.

Skill Drill 7-5: Donning a Level C Chemical-Protective Clothing Ensemble

Test your knowledge of this skill drill by placing the photos below in the correct order. Number the first step with a "1," the second step with a "2," and so on.

______ Don the inner gloves.

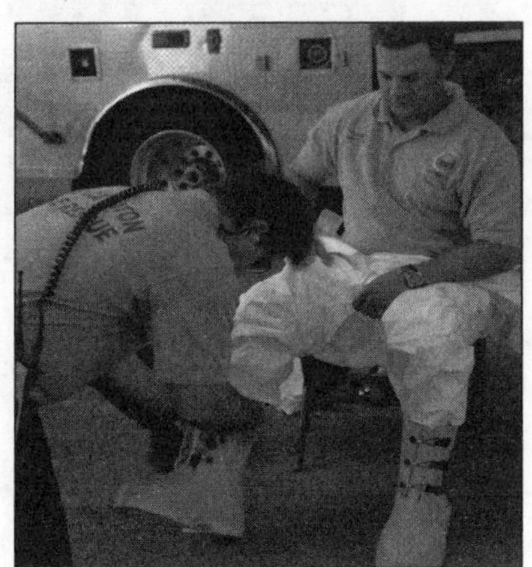

______ Conduct a pre-entry briefing, medical monitoring, and equipment inspection. While seated, pull on the suit to waist level; pull on the chemical boots over the top of the chemical suit. Pull the suit boot covers over the tops of the boots.

______ Don APR/PAPR. Pull the hood over the head and APR/PAPR face piece. Place the helmet on the head. Pull on the outer gloves. Review hand signals and indicate that you are okay.

______ With assistance, complete donning the suit by placing both arms in the suit and pulling the suit over the shoulders. Instruct the assistant to close the chemical suit by closing the zipper and sealing the splash flap.

Skill Drill 7-6: Doffing a Level C Chemical-Protective Clothing Ensemble

Test your knowledge of this skill drill by filling in the correct words in the photo captions.

1. After completing ______________, proceed to the clean area. As with Level B, the assistant opens the chemical splash flap and suit zipper. Remove the hands from the outer gloves and arms from the sleeves. Instruct the assistant to begin at the head and roll the suit down below waist level. Instruct the assistant to complete rolling down the suit and take the ______________ ______________ and suit away. The assistant helps remove the inner gloves. Remove APR/PAPR. Remove the helmet.

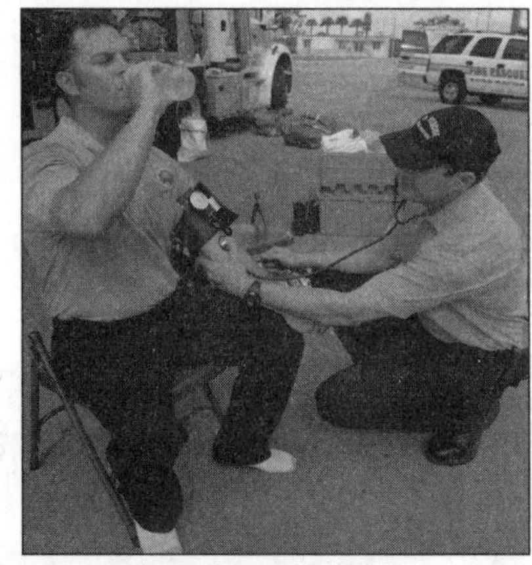

2. Go to the ______________ area for medical monitoring, rehydration, and personal decontamination shower.

Skill Drill 7-7: Donning a Level D Chemical-Protective Clothing Ensemble

Test your knowledge of this skill drill by filling in the correct words in the photo caption.

1. Conduct a pre-entry briefing and equipment inspection. Don the Level D suit. Don the boots. Don safety glasses or ________________ ________________. Don a hard hat. Don gloves, a face ________________, and any other required equipment.

Mission-Specific Competencies: Technical Decontamination

Workbook Activities

The following activities have been designed to help you. Your instructor may require you to complete some or all of these activities as a regular part of your hazardous materials training program. You are encouraged to complete any activity that your instructor does not assign as a way to enhance your learning in the classroom.

Chapter Review

The following exercises provide an opportunity to refresh your knowledge of this chapter.

Matching

Match each of the terms in the left column to the appropriate definition in the right column.

_____	**1.** Decontamination	**A.**	The process used to destroy recognized pathogenic microorganisms
_____	**2.** Chemical degradation	**B.**	The process of mixing a spongy material into a spilled liquid and picking up the mixture together
_____	**3.** Sterilization	**C.**	The process of chemically treating a hazardous liquid to turn it into a solid material, making the material easier to handle
_____	**4.** Absorption	**D.**	Occurs when a natural or artificial process causes the breakdown of a chemical substance
_____	**5.** Solidification	**E.**	The process of adding a material to a contaminant, which then adheres to the surface of the material for collection
_____	**6.** Disinfection	**F.**	The process of removing any form of contaminant from a person, an object, or the environment
_____	**7.** Adsorption	**G.**	A process using heat, chemical means, or radiation to kill all microorganisms
_____	**8.** Dilution	**H.**	The removal of dusts, particles, and some liquids by sucking them up into a container
_____	**9.** Evaporation	**I.**	Uses plain water or a soap-and-water mixture to lower the concentration of a hazardous material while flushing it off a contaminated person or object
_____	**10.** Vacuuming	**J.**	A natural form of chemical degradation that allows a chemical substance to stabilize without human intervention

Multiple Choice

Read each item carefully, and then select the best response.

_____ **1.** What is the situation called in which people, animals, or the environment come into direct contact with a hazardous material or the hazardous component of a weapon of mass destruction (WMD)?
- **A.** Contamination
- **B.** Dispersion
- **C.** Transference
- **D.** Integration

______ **2.** Which agency is responsible for laws governing the disposal of absorbent materials?
A. Fire department
B. Federal government
C. Department of Transportation
D. Emergency response team

______ **3.** Which method of decontamination is used during incidents involving unknown agents and large groups of people?
A. Emergency decontamination
B. Group decontamination
C. Gross decontamination
D. Mass decontamination

______ **4.** Which decontamination procedure mixes a spongy material with a liquid hazardous material?
A. Absorption
B. Adsorption
C. Dilution
D. Vapor dispersal

______ **5.** Which of the following is a two-step removal process for items that cannot be properly decontaminated?
A. Disinfection
B. Solidification
C. Isolation and disposal
D. Rapid mass decontamination

______ **6.** During decontamination, what is usually the last item of clothing removed?
A. Shoes
B. SCBA mask
C. Inner gloves
D. Face shield

______ **7.** Removed equipment should be placed:
A. near the entrance to the decontamination corridor.
B. in the hot zone.
C. in the cold zone.
D. in the hazardous materials truck.

______ **8.** After personnel are thoroughly decontaminated, they should proceed to:
A. the rehabilitation area.
B. EMS personnel.
C. the incident commander.
D. the Operations section.

______ **9.** When leaving the hot zone, all personal clothing should be:
A. diluted.
B. solidified.
C. placed in a drop area to be cleaned later.
D. burned.

______ **10.** Hazardous materials that have been neutralized have been:
A. diluted.
B. solidified.
C. bagged and tagged.
D. made safe by minimizing the corrosivity of an acid or base.

Labeling

Label the following diagrams with the correct terms.

1. Physical methods of technical decontamination.

A. ______________________________

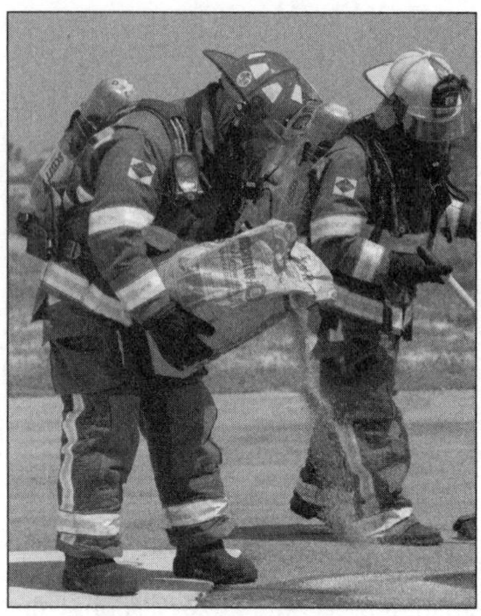

B. ______________________________

Vocabulary

Define the following terms using the space provided.

1. Decontamination team:

____

____

____

2. Contamination:

____

____

____

3. Adsorption:

____

____

____

4. Solidification:

____

____

____

5. Sterilization:

____

____

____

Fill-in

Read each item carefully, and then complete the statement by filling in the missing word(s).

1. The ______________ ______________ is a controlled area, usually within the warm zone, where decontamination procedures take place.
2. ______________ ______________ consists of a pre-wash that occurs before technical decontamination can take place.
3. When liquids with a high vapor pressure are spilled, responders may elect to take no action, but instead allow the substance to ______________.
4. The process used to destroy disease-causing microorganisms, excluding spores, is called ______________.
5. Fire fighters tend to use ______________ as the first decontamination method.
6. ______________ involves removing contaminated items from the primary incident site and storing them in a designated area; ______________ is the legal transportation of these items to approved facilities to be stored, incinerated, buried in a hazardous waste landfill, or otherwise handled.
7. The opposite of absorption is ______________.
8. ______________ ______________ is performed after gross decontamination and is a more thorough cleaning process.
9. Prior to any contaminated responders or victims passing through the decontamination corridor, the corridor can be considered ______________.
10. Whenever possible, ______________ the hazardous material before beginning decontamination.

True/False

If you believe the statement to be more true than false, write the letter "T" in the space provided. If you believe the statement to be more false than true, write the letter "F."

1. ______ Emergency medical responders are responsible for establishing a decontamination corridor for the initial emergency response crews and victims.
2. ______ During gross decontamination, hospital staff use low-pressure, high-volume water flow to rinse off and dilute contaminants.
3. ______ Vacuuming is the removal of dusts, particles, and some liquids by sucking them into a container.
4. ______ Personnel leaving the hot zone should place used tools in a tool drop area near the decontamination corridor.
5. ______ An item that will be submitted as evidence should arrive for decontamination double bagged.

Short Answer

Complete this section with short written answers using the space provided.

1. Identify and provide a brief description of the four major categories of decontamination.

Word Fun

The following crossword puzzle is an activity provided to reinforce correct spelling and understanding of terminology associated with hazardous materials. Use the clues provided to complete the puzzle.

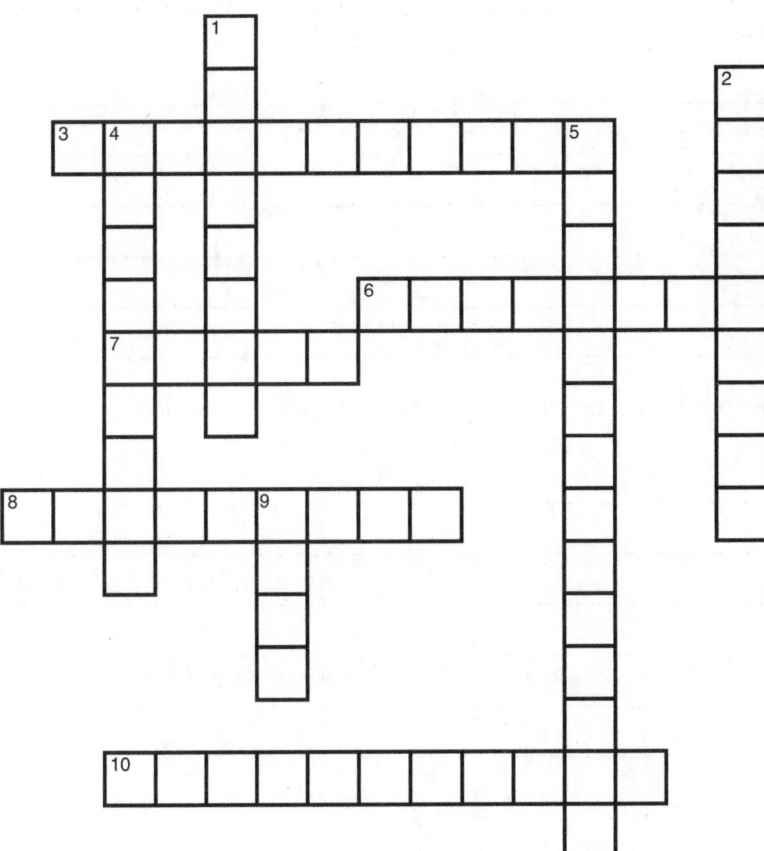

CLUES

Across

3 Chemical ______________ is a natural or artificial process that causes the breakdown of a chemical substance.

6 The process of adding some substance—usually water—in an attempt to weaken the concentration of another substance.

7 ______________ decontamination is a technique for significantly reducing the amount of surface contaminant by application of a continuous shower of water prior to the removal of outer clothing.

8 The process of cleaning up dusts, particles, and some liquids using a vacuum with high-efficiency particulate air (HEPA) filtration to prevent recontamination of the environment.

10 A natural form of chemical degradation in which a liquid material becomes a gas, allowing for dissipation of a liquid spill.

Down

1 The decontamination ______________ is a controlled area within the warm zone where decontamination takes place.

2 ______________ decontamination is a multistep process of carefully scrubbing and washing contaminants off of a person or object, collecting runoff water, and collecting and properly handling all items; it takes place after gross decontamination.

4 ______________ decontamination is the process of removing the bulk of contaminants off of a victim without regard for containment. It is used in potentially life-threatening situations, without the formal establishment of a decontamination corridor.

5 The method used when the corrosivity of an acid or a base needs to be minimized. This process accomplishes decontamination by way of a chemical reaction that alters the material's pH.

9 ______________ decontamination is the physical process of reducing or removing surface contaminants from large numbers of victims in potentially life-threatening situations in the fastest time possible.

Fire Alarms

The following real case scenarios will give you an opportunity to explore the concerns associated with hazardous materials. Read each scenario, and then answer each question in detail.

1. Your engine company is dispatched to a pesticide spill at a local hardware store. On arrival, you see several store patrons covered with liquid and powder. Your officer orders your company to set up emergency decontamination for the contaminated patrons. How will you proceed?

____

____

____

____

____

____

2. Your Lieutenant has given you the chance to prepare a short presentation on alternative decontamination procedures. Which topics will you discuss?

____

____

____

____

____

____

Skill Drills

Skill Drill 8-1: Performing Technical Decontamination on a Responder

Test your knowledge of this skill drill by placing the photos below in the correct order. Number the first step with a "1," the second step with a "2," and so on.

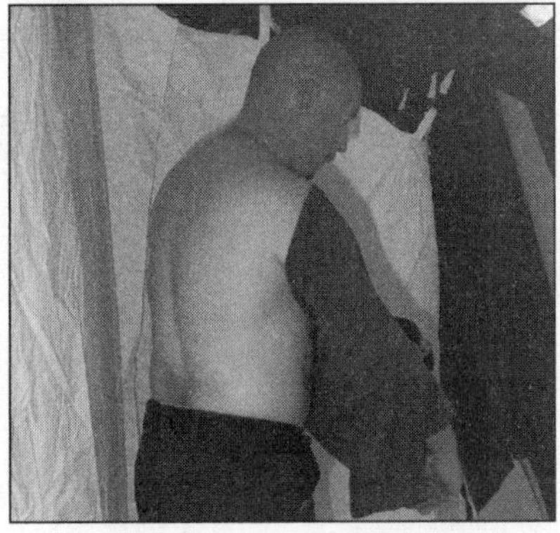

______ Remove personal clothing. Proceed to the rehabilitation area for medical monitoring, rehydration, and personal decontamination shower.

______ Perform gross decontamination, if necessary.

______ Perform technical decontamination. Wash and rinse responder one to three times.

______ Drop any tools and equipment.

______ Remove outer hazardous materials–protective clothing.

Mission-Specific Competencies: Mass Decontamination

Workbook Activities

The following activities have been designed to help you. Your instructor may require you to complete some or all of these activities as a regular part of your hazardous materials training program. You are encouraged to complete any activity that your instructor does not assign as a way to enhance your learning in the classroom.

Chapter Review

The following exercises provide an opportunity to refresh your knowledge of this chapter.

Matching

Match each of the terms in the left column to the appropriate definition in the right column.

_____	**1.** Ambulatory victims	**A.** The process of adding some substance—usually water—to a contaminant to decrease its concentration
_____	**2.** Washing	**B.** A process of removing contaminated items that cannot be properly decontaminated from the incident scene
_____	**3.** Dilution	**C.** Are able to walk on their own
_____	**4.** Isolation and disposal	**D.** Victims are doused with a simple soap-and-water solution and then fully rinsed with water
_____	**5.** *ERG*	**E.** Used as a resource for responders' initial actions

Multiple Choice

Read each item carefully, and then select the best response.

_____ **1.** Which of the following is not essential to ensuring the success of mass decontamination?
- **A.** Identify the contaminant if at all possible.
- **B.** Select and use the proper level of personal protective equipment.
- **C.** Have a predetermined process or procedure to perform decontamination.
- **D.** Assess the amount of property affected.

_____ **2.** Which of the following is not appropriate for mass decontamination?
- **A.** Fog-type nozzles attached to pumpers opposite each other
- **B.** Aerial ladder device providing an overhead spray pattern
- **C.** Deluge guns
- **D.** Prepackaged mass decontamination showers

_____ **3.** Separate areas for mass decontamination might be divided into which of the following categories?
- **A.** Ambulatory and nonambulatory victims
- **B.** Separate, gender-specific showering areas
- **C.** Both A and B are correct.
- **D.** None of the above; there is no time for separation during mass decontamination.

_____ **4.** Which technique is particularly suitable for dealing with contaminated personal clothing?
- **A.** Dilution
- **B.** Isolation
- **C.** Washing
- **D.** Disposal

_____ 5. The primary audience for the *ERG* includes all of the following, except:
  A. fire fighters.
  B. police officers.
  C. other emergency services personnel.
  D. truck drivers hauling hazardous materials.

Labeling

Label the following diagrams with the correct terms.

1. An example of a simple mass decontamination corridor using two fire engines.

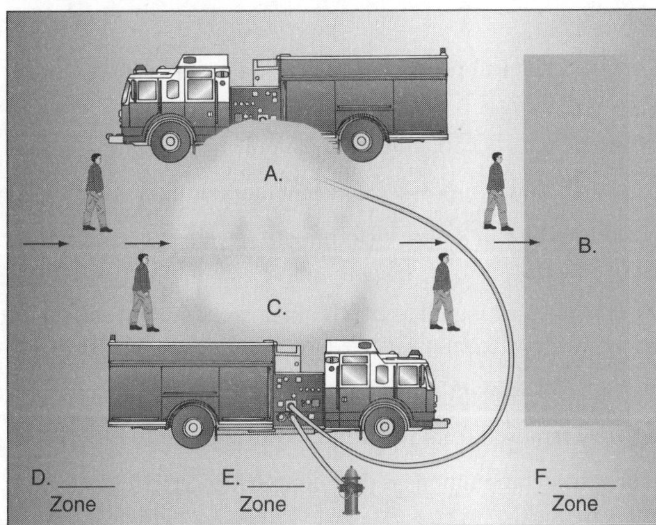

A. ______________________________

B. ______________________________

C. ______________________________

D. ______________________________

E. ______________________________

F. ______________________________

Vocabulary

Define the following terms using the space provided.

1. Mass decontamination:

____

____

____

2. Dilution:

____

____

____

3. Isolation and disposal:

____

____

____

Fill-in

Read each item carefully, and then complete the statement by filling in the missing word(s).

1. A water temperature of ______________ is ideal but may not be possible.
2. In most cases, significant medical treatment should be provided after ______________, in a designated medical treatment area.
3. When flushing the nonambulatory victims with water, take care to avoid compromising the victim's ______________ with water during the process.
4. Once the incident progresses past its first phase, generally beyond the first ______________ minutes, the *ERG* should not be used as a primary source of information.
5. When possible, make every attempt to track the ______________ and ______________ taken from the victims.

True/False

If you believe the statement to be more true than false, write the letter "T" in the space provided. If you believe the statement to be more false than true, write the letter "F."

1. ______ Mass decontamination and emergency decontamination are similar, except that emergency decontamination needs to be addressed much more quickly.
2. ______ Water washing will not change the physical or chemical properties of CO, so the application of water would yield minimum benefit.
3. ______ Controlling runoff is a primary objective when mass decontamination is implemented.
4. ______ Decontaminating ambulatory victims is a much slower process than decontaminating nonambulatory victims.
5. ______ When decontaminating nonambulatory victims, be sure to leave clothing or towels underneath the victim to allow for absorption of the product.
6. ______ Because water is a good general-purpose solvent, washing off as much of the contaminant as possible with a massive water spray is the best and quickest way to decontaminate a large group of people.
7. ______ Contaminated patients will almost always wait for responders to establish a formal decontamination area.
8. ______ Some viscous chemicals cannot be completely removed from the skin by washing alone.
9. ______ Evidence preservation is not an important consideration in mass decontamination activities.
10. ______ Naturally occurring barriers can be used to direct a moving group and create a manageable traffic flow.

Short Answer

Complete this section with short written answers using the space provided.

1. Describe the locations where the NFPA 704 system might be used.

____

____

____

____

2. Describe the locations where a placard system might be used.

____

____

____

____

Word Fun

The following crossword puzzle is an activity provided to reinforce correct spelling and understanding of terminology associated with hazardous materials. Use the clues provided to complete the puzzle.

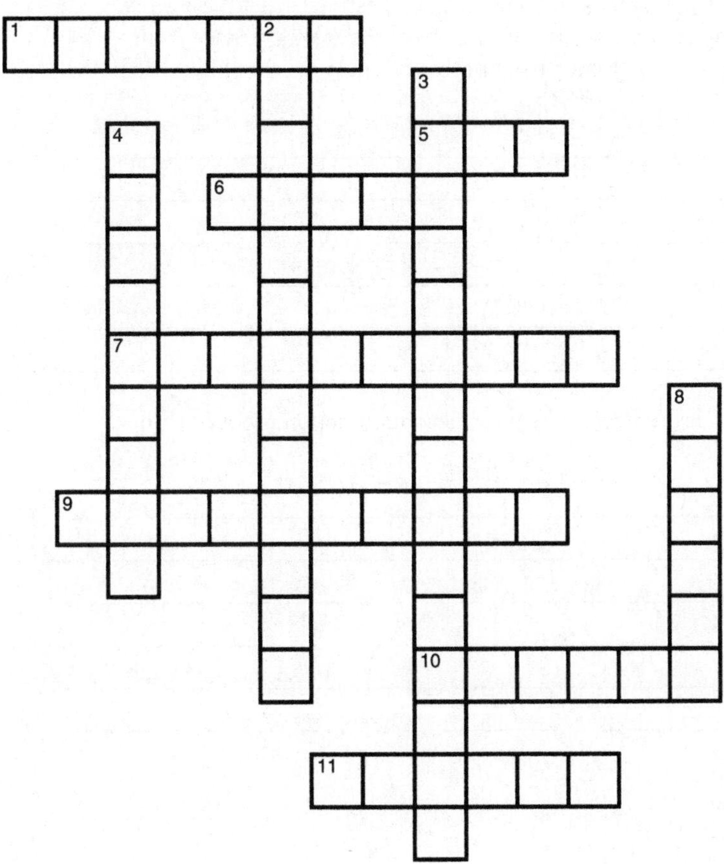

CLUES

Across

1 The process of dousing contaminated victims with a simple soap-and-water solution. The victims are then rinsed using water.

5 When a quick reference is required, the ______________ may prove useful for responders operating at a hazardous materials incident. (abbreviation)

6 When it comes to crowd control, remember that calmness and orderly behavior are just as contagious as ______________.

7 A(n) ______________ victim can walk on his or her own and can usually perform self rescue with direction and guidance from rescuers.

9 ______________ devices may be used to evaluate the completeness of decontamination.

10 Once ambulatory victims have been flushed with water inside the decontamination corridor, direct the contaminated victims to the ______________ area.

11 A(n) ______________ basket may be used to carry those victims in a hazardous materials incident who cannot walk.

Down

2 A(n) ______________ victim is one who cannot walk under his or her own power. These types of victims require more time and personnel when it comes to rescue.

3 Mass ______________ is the physical process of reducing or removing surface contaminants from large numbers of victims in potentially life-threatening situations in the fastest time possible.

4 ______________ and disposal is a two-step removal process for items that cannot be properly decontaminated.

8 Make sure to instruct ambulatory victims to remove their clothing ______________ entering any mass decontamination process.

Fire Alarms

The following real case scenarios will give you an opportunity to explore the concerns associated with hazardous materials. Read each scenario, and then answer each question in detail.

1. You have been dispatched to a university indoor sports arena where a large number of attendees have been exposed to an unknown agent that is severely irritating their skin. On arrival, you estimate that there are hundreds of anxious people with painful reddening of their skin streaming out of the main entrance. Your nearest mobile decontamination equipment will be coming from a neighboring town. What should you do in the interim?

____

____

____

____

____

____

2. University officials are expressing concern about the runoff from the mass decontamination process. Is this a concern?

____

____

____

____

____

____

Skill Drills

Skill Drill 9-1: Performing Mass Decontamination on Ambulatory Victims

Test your knowledge of this skill drill by filling in the correct words in the photo captions.

1. Ensure you have the appropriate PPE to protect against the chemical threat. Stay clear of the product and do not make physical contact with it. Make an effort to contain ______________ by directing victims out of the ______________ ______________ and into a suitable location.

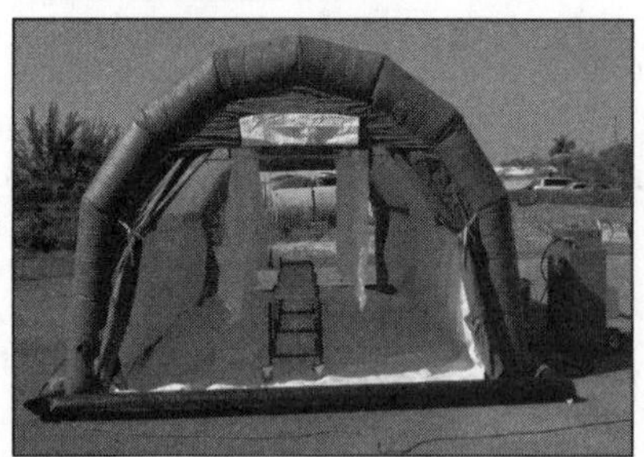

2. Set up the appropriate type of mass decontamination system based on the type of ______________, equipment, and/or system available.

3. Instruct victims to _______________ their contaminated clothing and walk through the decontamination corridor. Flush the contaminated victims with _______________.

4. Direct the contaminated victims to the _______________ area.

Skill Drill 9-2: Performing Mass Decontamination on Nonambulatory Victims

Test your knowledge of this skill drill by placing the photos below in the correct order. Number the first step with a "1," the second step with a "2," and so on.

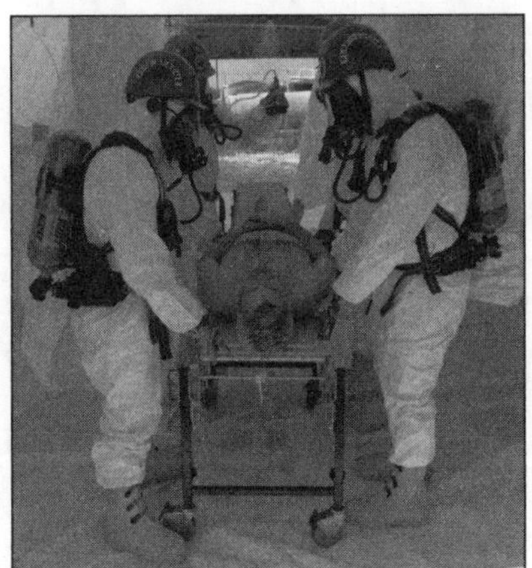

______ Flush the contaminated victims with water.

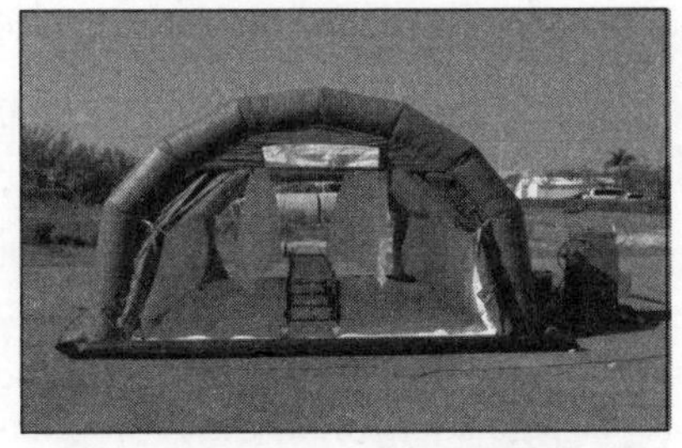

______ Set up the appropriate type of mass decontamination system based on the type of equipment available.

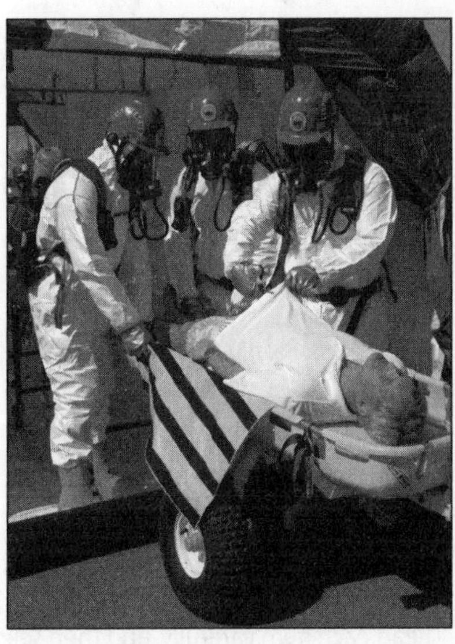

______ Ensure you have the appropriate PPE to protect against the chemical threat. Remove the victim's clothing. Do not leave any clothing underneath the victim; these items may wick the contamination to the victim's back and hold it there, potentially worsening the exposure.

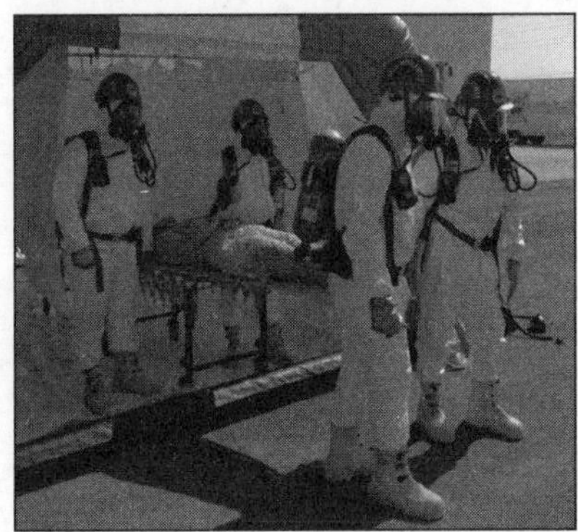

______ Move the victim to a designated triage area for medical evaluation.

Mission-Specific Competencies: Evidence Preservation and Sampling

Workbook Activities

The following activities have been designed to help you. Your instructor may require you to complete some or all of these activities as a regular part of your hazardous materials training program. You are encouraged to complete any activity that your instructor does not assign as a way to enhance your learning in the classroom.

Chapter Review

The following exercises provide an opportunity to refresh your knowledge of this chapter.

Matching

Match each of the terms in the left column to the appropriate definition in the right column.

_____ **1.** Evidence

_____ **2.** Investigative authority

_____ **3.** Chain of custody

_____ **4.** Trace evidence

_____ **5.** Demonstrative evidence

A. A suspect's clothing contains residue of the same ignitable liquid found at the scene of a fire

B. Refers to all of the information that is gathered and used by an investigator in determining the cause of an incident

C. The agency that has the legal jurisdiction to enforce a local, state, or federal law or regulation and is the appropriate law enforcement organization to investigate and prosecute

D. Anything that can be used to validate a theory to show how something could have occurred

E. Also known as chain of evidence or chain of possession

Multiple Choice

Read each item carefully, and then select the best response.

_____ **1.** Which of the following is responsible for investigating suspicious letters or packages involving hazardous materials that are sent through the mail?
A. Postal Inspection Service
B. Drug Enforcement Administration
C. Federal Bureau of Investigation
D. Environmental Protection Agency

_____ **2.** A minute quantity of physical evidence that is conveyed from one place to another is referred to as:
A. physical evidence.
B. evidence.
C. trace evidence.
D. demonstrative evidence.

_____ **3.** Burn patterns on a wall or an empty gasoline can left at the scene of an incident are examples of:
A. circumstantial evidence.
B. contaminated evidence.
C. investigative evidence.
D. physical evidence.

_____ **4.** Information that can be used to prove a theory based on facts that were observed firsthand is considered to be:

A. circumstantial evidence.
B. contaminated evidence.
C. investigative evidence.
D. physical evidence.

_____ **5.** Fences, excessive window coverings, or enhanced ventilation and air filtration systems may be indicators of:

A. evidence collection facilities.
B. illicit laboratories.
C. EOD facilities.
D. direct evidence.

Vocabulary

Define the following terms using the space provided.

1. Chain of custody:

____

____

____

2. Investigative authority:

____

____

____

3. Demonstrative evidence:

____

____

____

Fill-in

Read each item carefully, and then complete the statement by filling in the missing word(s).

1. ______________ evidence consists of items that can be observed, photographed, measured, collected, examined in a laboratory, and presented in court to prove or demonstrate a point.

2. Facts that can be observed or reported firsthand are referred to as ______________ evidence.

3. The process of protecting potential evidence until it can be documented, sampled, and collected appropriately is referred to as ______________ ______________.

4. Evidence that is altered from its original state in any way is considered to be _______________.

5. It is essential that whenever an explosive device is suspected to be involved, the appropriate _______________ _______________ _______________ personnel are notified.

6. _______________/_______________ records kept by the incident commander are a good way to document the identity and purpose of all personnel who enter an area that has been categorized as a crime scene.

7. Biological agents should be packaged only in _______________ and _______________ containers to ensure that cross-contamination with other microorganisms does not occur.

8. The use of a computer model to demonstrate how a fire could spread would be an example of _______________ evidence.

9. Hazardous materials incidents are often very complex and dynamic situations. It is not uncommon for a _______________ _______________, consisting of multiple disciplines and jurisdictional agencies, to be established at such events.

10. _______________ are often the first link in the chain of custody.

True/False

If you believe the statement to be more true than false, write the letter "T" in the space provided. If you believe the statement to be more false than true, write the letter "F."

1. ______ Regardless of the type of attack or weapon dissemination method, it is imperative that evidence be preserved, sampled, and collected properly.

2. ______ An intentional release or attack involving hazardous materials would be investigated by the Environmental Protection Agency.

3. ______ Methamphetamine is a psychostimulant drug manufactured illegally in illicit laboratories.

4. ______ Hazardous materials first responders at the scene are in the best position to decide whether the evidence they find will be admissible in court and worthy of preservation.

5. ______ Plastic containers should not be used to hold evidence-containing petroleum products because these chemicals may lead to deterioration of the plastic.

Short Answer

Complete this section with short written answers using the space provided.

1. Describe the 12-step process recommended by the FBI regarding the collection or sampling of evidence.

2. List several indicators that legitimate toxic industrial chemicals were released intentionally.

CLUES

Word Fun

The following crossword puzzle is an activity provided to reinforce correct spelling and understanding of terminology associated with hazardous materials. Use the clues provided to complete the puzzle.

Across

1 The ______________ Service is the federal agency charged with securing and managing the U.S. mail system. (2 words)

3 ______________ personnel are trained to detect, identify, evaluate, render safe, recover, and dispose of unexploded explosive devices. (abbreviation)

5 ______________ evidence are materials used to explain a theory or an event.

6 A(n) ______________ basket may be used to carry those victims in a hazardous materials incident who cannot walk.

8 A psychostimulant drug manufactured illegally in illicit laboratories.

9 The ______________ authority has the legal jurisdiction to enforce a local, state, or federal law or regulation; the most appropriate law enforcement organization to ensure the successful analysis and prosecution of a case.

Down

1 ______________ evidence can be observed, photographed, measured, collected, examined in a laboratory, and presented in court to prove or demonstrate a point.

2 ______________ evidence is a minute quantity of physical evidence that is conveyed from one place to another.

3 The ______________ Protection Agency, established in 1970, is charged with ensuring safe manufacturing, use, transportation, and disposal of hazardous substances.

4 Evidence ______________ is the process of protecting potential evidence until it can be documented, sampled, and collected appropriately.

6 Evidence ______________ is the process of collecting portions of a hazardous material/WMD for the purposes of field screening, laboratory testing, and, ultimately, criminal prosecution.

7 Information that is gathered and used by an investigator in determining the cause of an incident.

Fire Alarms

The following real case scenarios will give you an opportunity to explore the concerns associated with hazardous materials. Read each scenario, and then answer each question in detail.

1. You have been dispatched to a fire alarm in a residential community. On your arrival, a police officer indicates that there is no fire, but he states that neighbors suspect a clandestine laboratory is operating there. What indicators should you look for to consider this risk?

2. Your size-up of the area gives you the impression that the operators of this illegal laboratory may also have been disposing of hazardous wastes in the back yard of the property. What indicators should you look for to consider this risk?

Skill Drills

Skill Drill 10-1: Collecting and Processing Evidence

Test your knowledge of this skill drill by placing the photos below in the correct order. Number the first step with a "1," the second step with a "2," and so on.

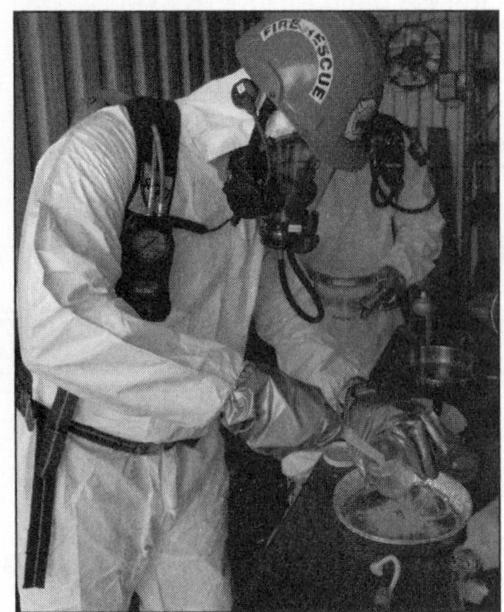

______ Place the evidence in the appropriate container.

______ Tag the evidence with labels.

______ Take photographs of the evidence.

______ Sketch, mark, and label the location of the evidence.

______ Document your findings.

Skill Drill 10-2: Securing, Characterizing, and Preserving the Scene

Test your knowledge of this skill drill by filling in the correct words in the photo captions.

1. Observe the scene for certain characteristics that could lead to the discovery of _______________.

2. Place caution tape around the scene to limit _______________.

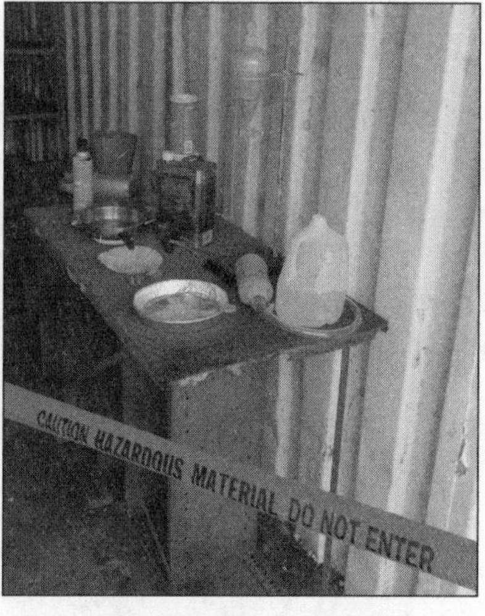

3. _______________ suspected evidence by protecting it from being disturbed.

Skill Drill 10-6: Collecting Samples Utilizing Equipment and Preventing Secondary Contamination

Test your knowledge of this skill drill by filling in the correct words in the photo captions.

1. The sampler obtains two samples: one sample to use for _______________ _______________ and another sample to preserve as _______________.

2. The _______________ holds open the evidence package or container so the sampler can place the evidence inside without cross-contaminating the evidence.

Skill Drill 10-7: Documenting Evidence

Test your knowledge of this skill drill by filling in the correct words in the photo captions.

1. The _______________ photographs and/or videotapes the sampling and collection process.

2. The documenter makes notes about the name of the person _______________ or _______________ the evidence, the physical location of the agent, the state of the agent, the _______________ present, the time of the sample collection, and the size and _______________ of the container.

Skill Drill 10-8: Evidence Labeling, Packaging, and Decontamination

Test your knowledge of this skill drill by filling in the correct words in the photo captions.

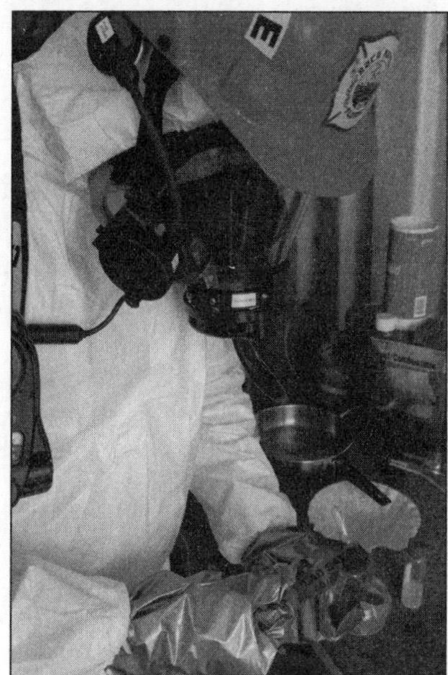

1. Seal the initial container with tape, and place your ______________ on the tape or seal. This step will prevent ______________.

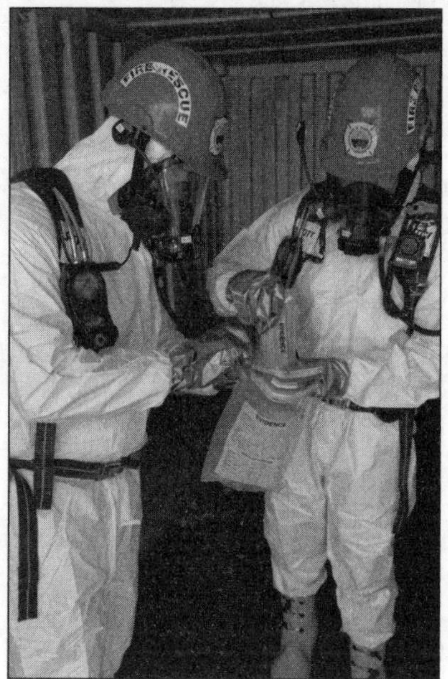

2. Place the initial container in a ______________ container and label it with a unique ______________ number, the name of the person who collected the item, and the location, time, and date of the evidence collection.

Mission-Specific Competencies: Product Control

Workbook Activities

The following activities have been designed to help you. Your instructor may require you to complete some or all of these activities as a regular part of your hazardous materials training program. You are encouraged to complete any activity that your instructor does not assign as a way to enhance your learning in the classroom.

Chapter Review

The following exercises provide an opportunity to refresh your knowledge of this chapter.

Matching

Match each of the terms in the left column to the appropriate definition in the right column.

	Term	Definition
______	1. Vapor dispersion	A. The process of keeping a hazardous material on the site or within the immediate area of the release
______	2. Confinement	B. Actions relating to stopping a hazardous materials container from leaking, such as patching, plugging, or righting the container
______	3. Containment	C. The process of lowering the concentration of vapors by spreading them out
______	4. Diversion	D. The process of controlling vapors by covering the product with foam or by reducing the temperature of the material
______	5. Dilution	E. The process of adding some substance to a product to weaken its concentration
______	6. Vapor suppression	F. Redirecting spilled material to an area where it will have less impact
______	7. Retention	G. The process of applying a material that will soak up and hold the hazardous material
______	8. Diking	H. Used when a liquid is flowing in a natural channel or depression, and its progress can be stopped by blocking the channel
______	9. Damming	I. The placement of impervious materials to form a barrier that will keep a hazardous material in liquid form from entering an area
______	10. Absorption	J. The process of creating an area to hold hazardous materials

Multiple Choice

Read each item carefully, and then select the best response.

______ 1. The process whereby a spongy material or spill pads are used to soak up a liquid hazardous material is known as:
- A. absorption.
- B. adsorption.
- C. diversion.
- D. retention.

______ 2. The process of redirecting the flow of a liquid away from an endangered area to an area where it will have less impact is known as:
- A. retention.
- B. repulsion.
- C. diversion.
- D. diking.

______ 3. The phase of a hazardous materials incident when the imminent danger to people, property, and the environment has passed or is controlled is referred to as:
A. mitigation.
B. recovery.
C. response.
D. suppression.

______ 4. Foam that is utilized when large volumes of foam are required for spills or fires in warehouses, tank farms, and hazardous waste facilities is known as:
A. high-expansion foam.
B. protein foam.
C. fluoroprotein foam.
D. aqueous film-forming foam.

______ 5. A containment technique in which a dam is placed across a small stream or ditch to completely stop the flow of materials through the channel is considered a(n):
A. overflow dam.
B. underflow dam.
C. Wheatstone bridge.
D. complete dam.

______ 6. The process of attempting to keep the hazardous material on the site or within the immediate area of the release is known as:
A. confinement.
B. containment.
C. exposure.
D. suppression.

______ 7. Most flammable and combustible liquid fires can be extinguished by the use of:
A. water.
B. carbon monoxide.
C. foam.
D. dilution.

______ 8. As fire fighters approach a hazardous materials incident, they should look for:
A. a means of egress.
B. damage to property or surfaces.
C. natural control points.
D. plumes of smoke.

______ 9. Why is the technique of absorption difficult for operational personnel to implement?
A. It creates an extensive clean-up process.
B. It requires the appropriate material matches.
C. It involves a large number of personnel.
D. It generally involves being in close proximity to the spill.

______ 10. The process of creating an area to hold hazardous materials is called:
A. retention.
B. diking.
C. damming.
D. diversion.

_____ 11. The addition of another liquid to weaken the concentration of a hazardous material is called:
A. dispersion.
B. dilution.
C. extension.
D. liquidation.

_____ 12. The process of lowering the concentration of vapors by spreading them out is called:
A. vapor suppression.
B. vapor release.
C. vapor evacuation.
D. vapor dispersion.

_____ 13. When the imminent danger has passed and clean-up and the return to normalcy have begun, the incident has reached the:
A. debriefing phase.
B. clean-up phase.
C. recovery phase.
D. termination phase.

_____ 14. Which phase of the incident includes the compilation of all records necessary for documentation of the incident?
A. Administration phase
B. Recovery phase
C. Wrap-up phase
D. Size-up phase

_____ 15. Who makes the decision to terminate a hazardous materials incident?
A. Safety officer
B. Incident commander
C. Operations officer
D. Planning officer

Vocabulary

Define the following terms using the space provided.

1. Exposures:

____

____

____

2. Underflow dam:

____

____

____

3. Recovery phase:

____

____

____

Fill-in

Read each item carefully, and then complete the statement by filling in the missing word(s).

1. The process of separating and diminishing harmful vapors is known as ______________ ______________.
2. A water spray is commonly used to ______________ vapors.
3. Highly volatile flammable liquids may be left to ______________ on their own without taking offensive action to clean them up.
4. If fire fighters expose themselves to ______________ risk, injury, exposure, or contamination, they only complicate the problem.
5. Fixed ammonia systems provide a good example of the effectiveness of using ______________ ______________.
6. ______________ is the process of creating an area to hold hazardous materials.
7. A(n) ______________ ______________ is placed across a small stream or ditch to completely stop the flow of materials through the channel.
8. Alcohol-resistant concentrates are formulated so that polar solvents will not ______________ the foam.
9. Most flammable and combustible ______________ ______________ can be extinguished by the use of foam.
10. ______________ refers to actions that stop the hazardous material from leaking or escaping its container.

True/False

If you believe the statement to be more true than false, write the letter "T" in the space provided. If you believe the statement to be more false than true, write the letter "F."

1. ______ In a hazardous materials incident, all emergency response personnel must first recognize and identify which hazardous materials may be present.
2. ______ The opposite of absorption is adsorption.
3. ______ All exposures need to be protected in the same way.
4. ______ Firefighting foams should be sprayed directly on the burning material and surface.
5. ______ In some cases the incident commander may decide to withdraw to a safe distance and let the hazardous materials incident run its course.
6. ______ The recovery phase and clean-up will likely require amounts of resources and equipment that are far beyond the capabilities of local responders.
7. ______ MC 307/DOT 407 cargo tanks are certified to carry chemicals that are transported at high pressure.
8. ______ Many chemical processes, or piped systems that carry chemicals, have a way to remotely shut down a system or isolate a valve.
9. ______ Dilution can be used only when the identity and properties of the hazardous material are known with certainty.
10. ______ A retention technique is used to redirect the flow of a liquid away from an area.

Short Answer

Complete this section with short written answers using the space provided.

1. List three types of firefighting foams.

____

____

____

CLUES

Word Fun

The following crossword puzzle is an activity provided to reinforce correct spelling and understanding of terminology associated with hazardous materials. Use the clues provided to complete the puzzle.

Across

2 The product-control process used when liquid is flowing in a natural channel or depression, and its progress can be stopped by constructing a barrier to block the flow.

3 The process in which a contaminant adheres to the surface of an added material—such as silica or activated carbon—rather than combining with it.

8 ______________ foam is a blended organic and synthetic foam that is made from animal by-products and synthetic surfactants.

10 ______________ foam is created by pumping large volumes of air through a small screen coated with a foam solution. (2 words)

11 The process by which people, animals, the environment, and equipment are subjected to or come into contact with a hazardous material.

Down

1 The placement of materials to form a barrier that will keep a hazardous material in liquid form from entering an area or that will hold the material in an area.

2 Vapor ______________ lowers the concentration of vapors by spreading them out, typically with a water fog from a hose line.

4 ______________ shut-off may be found at fixed facilities utilizing chemical processes or piped systems that carry chemicals. (2 words)

5 The process of purposefully collecting hazardous materials in a defined area.

6 The process of keeping a hazardous material within the immediate area of the release.

7 A(n) ______________ basket may be used to carry those victims in a hazardous materials incident who cannot walk.

9 The ______________ phase of a hazardous materials incident occurs after imminent danger has passed, when clean-up and the return to normalcy have begun.

Fire Alarms

The following real case scenarios will give you an opportunity to explore the concerns associated with hazardous materials. Read each scenario, and then answer each question in detail.

1. You have been dispatched to a diesel fuel spill on the state route south of town. On arrival at the site, you find 10 gallons of diesel fuel spilled across the roadway. Your company officer directs you to absorb the fuel with on-board absorbent. How will you absorb the spilled hazardous material?

2. Before you can begin absorbing the spilled diesel fuel, the combustible material is ignited by a road flare placed too close to the spill.

a. What can you use to extinguish the burning combustible fuel?

b. How should the extinguishing agent be applied?

Skill Drills

Skill Drill 11-1: Using Absorption/Adsorption to Manage a Hazardous Materials Incident

Test your knowledge of this skill by filling in the correct words in the photo captions.

1. Decide which _______________ is best suited for use with the spilled product. Access the location of the spill and stay clear of any spilled product.

2. Use detection and monitoring _______________ as well as reference sources to identify the product. Apply the appropriate material to control the spilled product.

3. Maintain _______________ of the absorbent/adsorbent materials and take appropriate steps for their disposal.

Skill Drill 11-4: Constructing a Dike

Test your knowledge of this skill by filling in the correct words in the photo captions.

1. Determine the best location for the dike. If necessary, dig a depression in the ground 6" to 8" (15 cm to 20 cm) deep. Ensure that plastic will not _______________ adversely with the spilled chemical. Use plastic to line the bottom of the depression, and allow for sufficient plastic to cover the dike wall.

2. Build a short wall with _______________ or other available materials.

3. Complete the dike installation, and ensure that its _______________ will contain the spilled product.

Skill Drill 11-8: Using Vapor Dispersion to Manage a Hazardous Materials Incident

Test your knowledge of this skill by filling in the correct words in the photo captions.

1. Determine the viability of a dispersion operation. Use the appropriate monitoring instrument to determine the boundaries of a safe work area. Ensure that _______________ _______________ in the areas have been removed or controlled.

2. Apply water from a distance to disperse vapors. _______________ the environment until the vapors have been adequately dispersed.

Skill Drill 11-12: Performing the Bounce-Off Method of Applying Foam

Test your knowledge of this skill by filling in the correct words in the photo captions.

1. Open the nozzle and _______________ to ensure that foam is being produced.

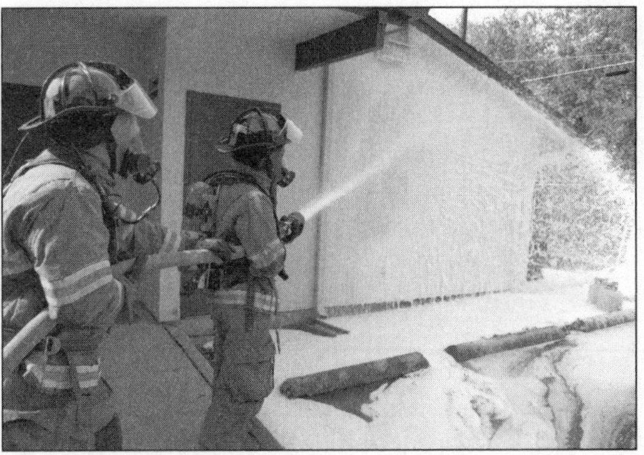

2. Move within a safe range of the product, and open the nozzle. Direct the stream of foam onto a solid structure, such as a wall or metal tank, so that the foam is directed off the object and onto the pool of product. Allow the foam to flow across the top of the pool of product until it is completely covered. Be aware that the foam may need to be banked off of several areas of the solid object so as to _______________ the burning product.

Mission-Specific Competencies: Victim Rescue and Recovery

Workbook Activities

The following activities have been designed to help you. Your instructor may require you to complete some or all of these activities as a regular part of your hazardous materials training program. You are encouraged to complete any activity that your instructor does not assign as a way to enhance your learning in the classroom.

Chapter Review

The following exercises provide an opportunity to refresh your knowledge of this chapter.

Matching

Match each of the terms in the left column to the appropriate definition in the right column.

_____	**1.** Ambulatory victims	**A.** A crew of fully qualified and equipped responders who are assigned to enter the hot zone
_____	**2.** Shelter-in-place	**B.** Individuals who function as a stand-by rescue crew of relief for those entering the hot zone
_____	**3.** Entry team	**C.** Are able to walk on their own
_____	**4.** Backup team	**D.** Sometimes this option is preferable to removing victims from a building
_____	**5.** START	**E.** A triage system for large-scale, mass-casualty incidents

Multiple Choice

Read each item carefully, and then select the best response.

_____ **1.** Which of the following represents a shift in mindset and tactics whereby the incident response reflects the fact that there is no chance of rescuing live victims?
- **A.** Contacting the property owners
- **B.** Response mode
- **C.** Recovery mode
- **D.** Triage mode

_____ **2.** When beginning a rescue in a hazardous materials situation, all of the following are required, except:
- **A.** an entry team.
- **B.** a backup team.
- **C.** decontamination person(s).
- **D.** a staging team.

_____ **3.** Which of the following techniques would not be used for a victim who is conscious and responsive but incapable of standing or walking?
- **A.** Two-person extremity carry
- **B.** Two-person seat carry
- **C.** Two-person chair carry
- **D.** Two-person walking assist

______ **4.** Which technique is particularly suitable when a victim must be carried through doorways, along narrow corridors, or up and down stairs?
  A. Two-person chair carry
  B. Two-person extremity carry
  C. Clothes drag
  D. Standing drag

______ **5.** In making the decision to act at a hazardous materials incident, the first decision concerns:
  A. the amount of property affected.
  B. a reasonable expectation of a positive outcome.
  C. the number of people involved.
  D. the type of material involved.

Labeling

Label the following diagrams with the correct terms.

1. Search and rescue/recovery equipment.

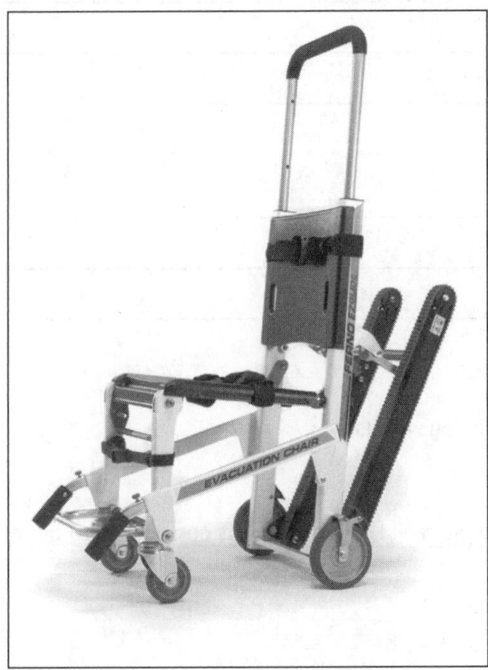

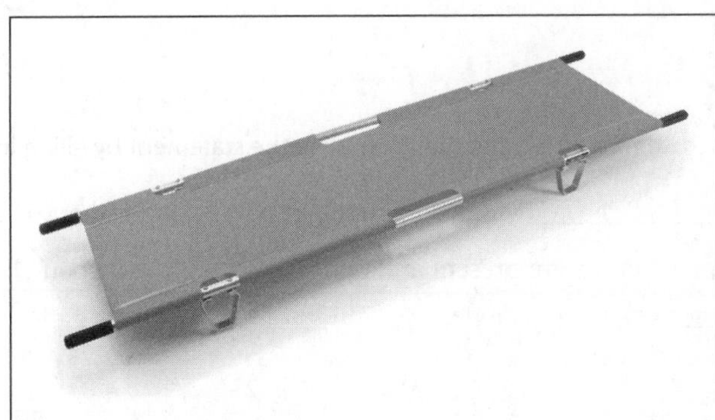

A. ______________________

B. ______________________

C. ______________________

Vocabulary

Define the following terms using the space provided.

1. Rescue mode:

____

____

____

2. Backup team:

____

____

____

3. Triage:

____

____

____

Fill-in

Read each item carefully, and then complete the statement by filling in the missing word(s).

1. ______________ victims are unable to walk under their own power.
2. When victims are present and determined to have a good chance for survival, the incident is considered to be in ______________ mode.

3. If there are six responders who will make up the entry team, there should be ______________ responders on the backup team.

4. The ______________-______________ ______________ ______________, also known as the sit pick, requires no equipment and can be performed in tight or narrow spaces.

5. The ______________ ______________ is used to move a victim who is on the floor or the ground and is too heavy for one responder to lift and carry alone.

6. The ______________ sling drag ensures the responder has a secure grip around the upper part of a victim's body, allowing for a quick and efficient exit from the dangerous area.

7. Whenever possible, a ______________ ______________ should be used to remove a victim from a vehicle.

8. OSHA requires that responders working in the hot zone work in ______________-member teams.

9. Nonambulatory victims in mass-casualty situations must be sorted out based on their medical priority using the ______________ method.

10. Victim search is not ______________ if a hazardous material exposure is not survivable.

True/False

If you believe the statement to be more true than false, write the letter "T" in the space provided. If you believe the statement to be more false than true, write the letter "F."

1. ______ In most hazardous materials incidents, at least five trained responders are required to make a rescue attempt.

2. ______ In hazardous materials response, the entry team/backup team ratio is always 1:1.

3. ______ Backup team members are dressed one PPE level down from that worn by the entry team.

4. ______ Typically, definitive medical care is rendered to victims during rescue mode.

5. ______ Emergency decontamination is performed in potentially life-threatening situations to rapidly remove the bulk of the contamination from an individual.

6. ______ Ambulatory victims who are within the line of sight and are able to walk may be directed and encouraged to leave the area under their own power.

7. ______ The cradle-in-arms carry can be used by one responder to carry a child or small adult.

8. ______ The best method for one person to remove a victim from a vehicle without compromising the neck and spine is with a long backboard rescue technique.

9. ______ Hazardous materials teams practiced the concept of rapid intervention before it was adopted in structural firefighting.

10. ______ Chemical gloves improve dexterity and make both clothing and skin easier to grasp.

Short Answer

Complete this section with short written answers using the space provided.

1. List the six emergency drags that can be used to remove unconscious victims from a dangerous situation.

2. Identify the acronym used in START triage, and describe the process.

Word Fun

The following crossword puzzle is an activity provided to reinforce correct spelling and understanding of terminology associated with hazardous materials. Use the clues provided to complete the puzzle.

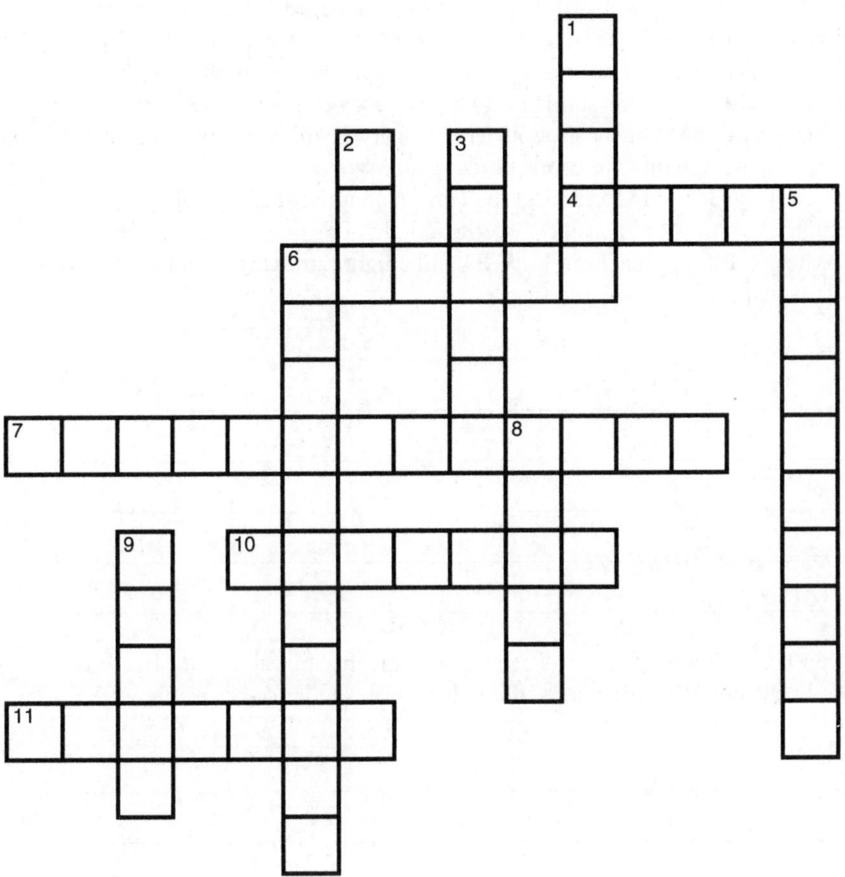

CLUES

Across

4 ________________ attempt to make a rescue alone.

6 Responders should attempt to understand the physical and chemical properties of a substance ________________ beginning any form of decontamination.

7 The one-person ________________ can be used if a victim is capable of walking. (2 words)

10 Typically, definitive ________________ care is not rendered to victims during rescue mode.

11 A(n) ________________ search is a rapidly conducted search, focusing on those victims who are easily seen, who are lightly trapped or disoriented, or who are otherwise found in easily accessible locations.

Down

1 A(n) ________________ board may be necessary to perform search, rescue, and recovery tasks, depending on the specific situation.

2 ________________ backup entry team member(s) should be provided for each entry team member.

3 A(n) ________________ basket may be used to carry those victims in a hazardous materials incident who cannot walk.

5 Those activities directed at locating endangered persons at an emergency incident, removing those persons from danger, treating injured victims, and providing for transport to an appropriate healthcare facility. (2 words)

6 This procedure requires the use of a large sheet, blanket, curtain, or rug; used to move a victim. (2 words)

8 A well-recognized system that provides several steps by which to conduct triage at large-scale mass-casualty incidents. (abbreviation)

9 During search, rescue, and recovery operations, at least one member of each search team should be equipped with a portable ________________.

Fire Alarms

The following real case scenarios will give you an opportunity to explore the concerns associated with hazardous materials. Read each scenario, and then answer each question in detail.

1. You are the officer in charge of a fire company responding to a report of a chemical leak at a manufacturing plant adjacent to an expressway. The alarm was called in by a contract security guard who detected a sharp, pungent odor and found an outdoor 10,000-gallon chemical storage tank leaking into a secondary containment. On your arrival onto the premises, you note that there is no placard on the container; however, the ID number is clearly stenciled and visible from a distance. Because the material is spilling into a secondary containment, there is no immediate threat of the materials entering storm drains or the ground. The security guard called in from a cell phone, but does not answer when your dispatcher attempts to contact him again for details. In his initial report, however, he stated that the material involved was Methyl Isocyanate. You can see that the security vehicle is parked alongside the tank and has its engine running and its yellow overhead lights flashing, and it appears that someone is lying on the ground next to the car. It is approaching 1:00 AM and a light rain has begun to fall. How can you safely rescue the downed security guard?

2. Your first ladder company reports that a second security guard has been found unconscious at a guard booth at the perimeter of the property. She is not breathing. What first aid actions should the ladder company crew take while they await an ambulance?

Skill Drills

Skill Drill 12-1: Performing a One-Person Walking Assist

Test your knowledge of this skill drill by filling in the correct words in the photo captions.

1. Help the victim to ________________.

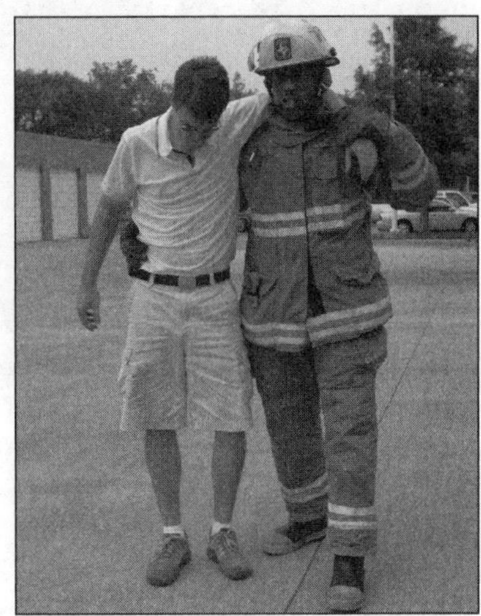

2. Have the victim place his or her arm around your neck, and hold onto the victim's ______________, which should be draped over your shoulder. Put your free arm around the victim's ______________ and help the victim to walk.

Skill Drill 12-2: Performing a Two-Person Walking Assist

Test your knowledge of this skill drill by placing the photos below in the correct order. Number the first step with a "1," the second step with a "2," and so on.

______ Once the victim is fully upright, drape the victim's arms around the necks and over the shoulders of the responders, each of whom holds one of the victim's wrists.

______ Assist walking at the victim's speed.

______ Two responders stand facing the victim, one on each side of the victim.

______ Both responders put their free arm around the victim's waist, grasping each other's wrists for support and locking their arms together behind the victim.

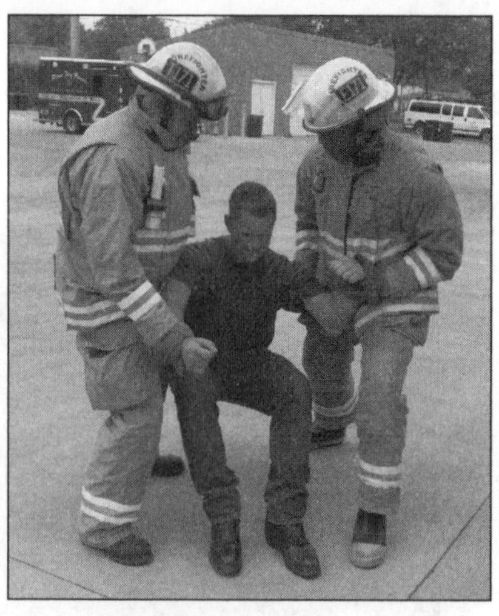

______ The responders assist the victim to a standing position.

Skill Drill 12-3: Performing a Two-Person Extremity Carry

Test your knowledge of this skill drill by placing the photos below in the correct order. Number the first step with a "1," the second step with a "2," and so on.

______ The second responder backs in between the victim's legs, reaches around, and grasps the victim behind the knees.

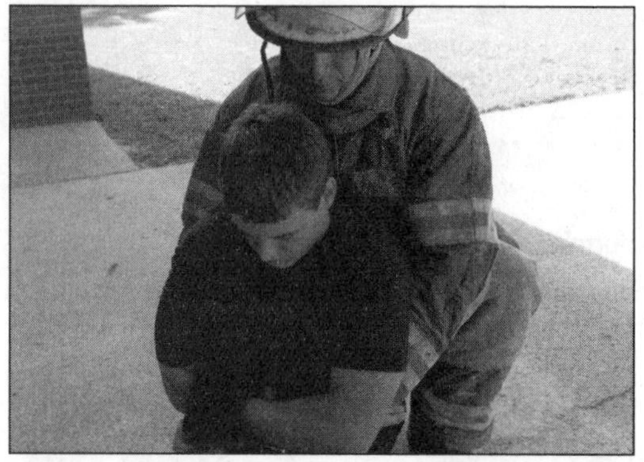

______ The first responder kneels behind the victim, reaches under the victim's arms, and grasps the victim's wrists.

______ The first responder gives the command to stand and carry the victim away, walking straight ahead. Both responders must coordinate their movements.

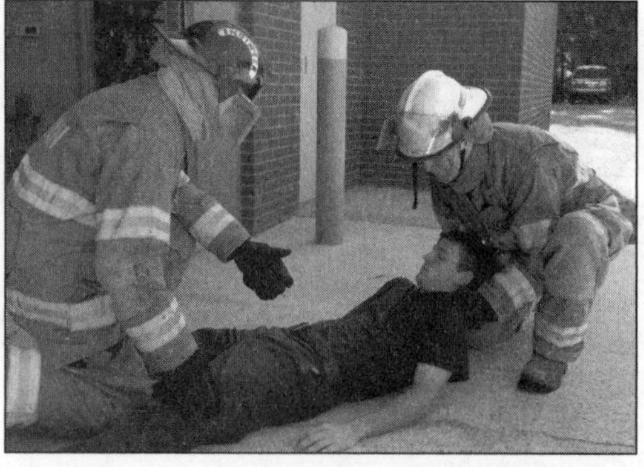

______ Two responders help the victim to sit up.

Skill Drill 12-4: Performing a Two-Person Seat Carry

Test your knowledge of this skill drill by placing the photos below in the correct order. Number the first step with a "1," the second step with a "2," and so on.

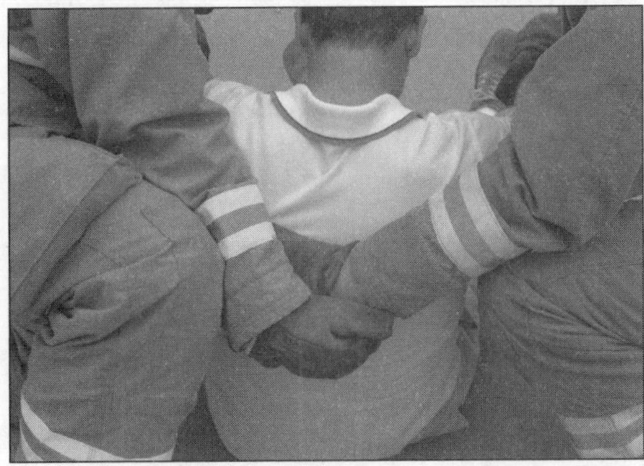

_______ Raise the victim to a sitting position and link arms behind the victim's back.

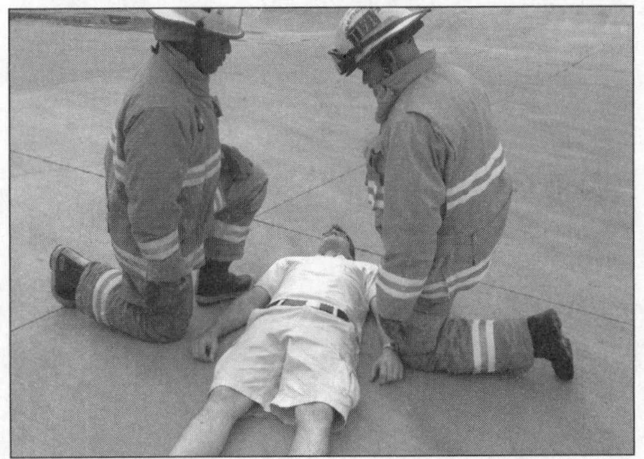

_______ Kneel beside the victim near the victim's hips.

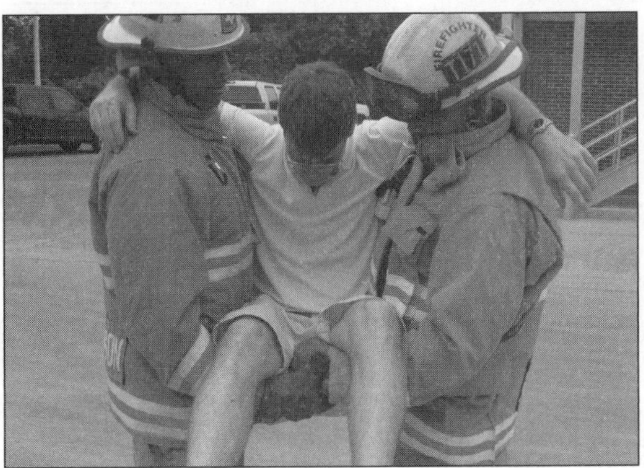

_______ If possible, the victim puts his or her arms around your necks for additional support.

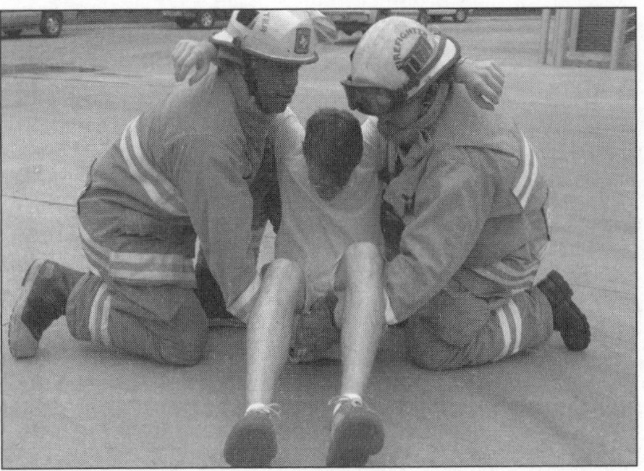

_______ Place your free arms under the victim's knees and link arms.

Skill Drill 12-5: Performing a Two-Person Chair Carry

Test your knowledge of this skill drill by filling in the correct words in the photo captions.

1. One responder stands behind the seated victim, reaches down, and grasps the ______________ of the chair.

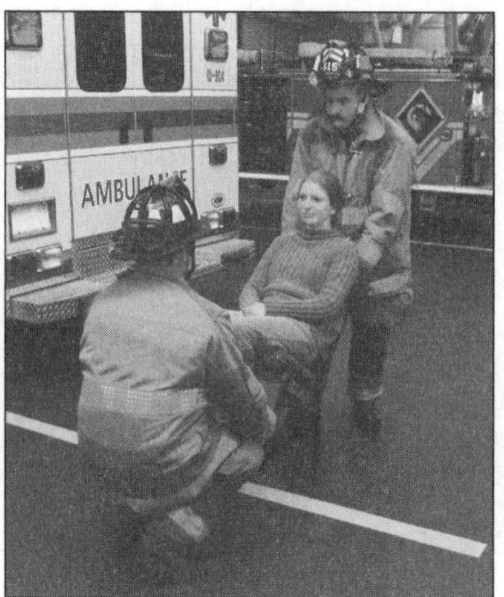

2. The responder tilts the chair slightly backward on its rear legs so that the second responder can step between the legs of the chair and grasp the tips of the chair's front legs. The victim's ______________ should be ______________ the legs of the chair.

3. When both responders are correctly positioned, the responder behind the chair gives the command to ______________ and walk away. Because the chair carry may force the victim's ______________ forward, watch the victim for ______________ problems.

Skill Drill 12-6: Performing a Cradle-in-Arms Carry

Test your knowledge of this skill drill by filling in the correct words in the photo captions.

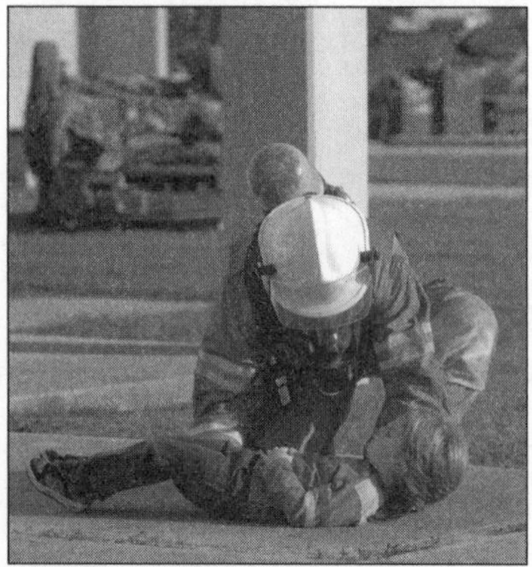

1. Kneel beside the child, and place one arm around the child's ______________ and the other arm under the ______________.

2. Lift slightly and roll the child into the hollow formed by your ______________ and ______________.

3. Be sure to use your ______________ muscles to stand.

Skill Drill 12-7: Performing a Clothes Drag

Test your knowledge of this skill drill by filling in the correct words in the photo captions.

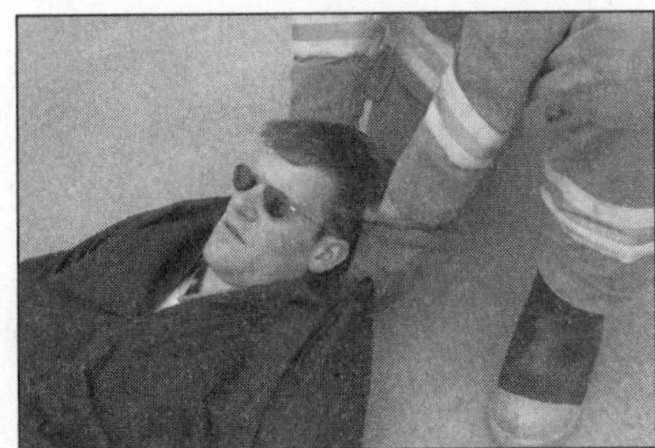

1. Crouch behind the victim's head, and grab the shirt or jacket around the collar and _______________ area.

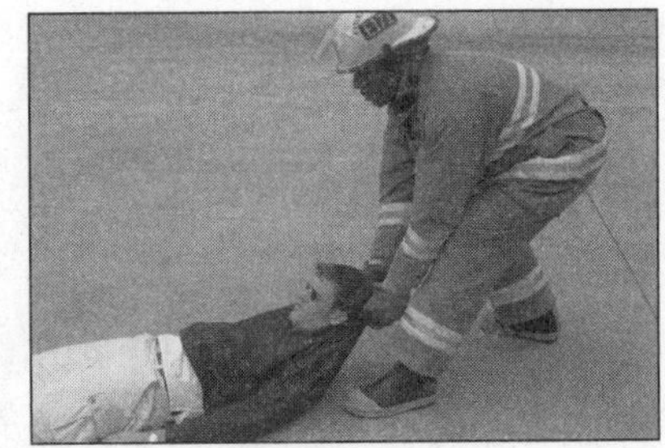

2. Lift with your _______________ until you are fully upright. Walk _______________, dragging the victim to safety.

Skill Drill 12-8: Performing a Blanket Drag

Test your knowledge of this skill drill by filling in the correct words in the photo captions.

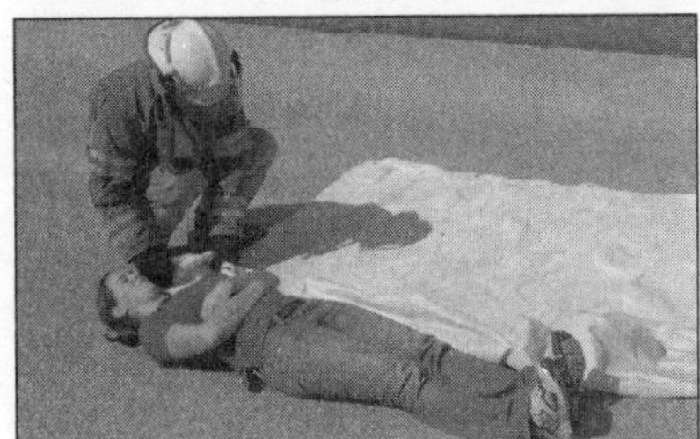

1. Stretch out the _______________ you are using next to the victim.

2. Roll the victim onto one side. Neatly bunch one _______________ of the material against the victim's body.

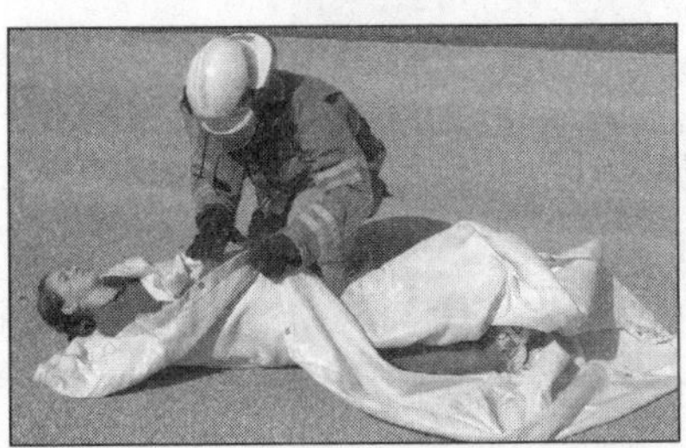

3. Lay the victim back down (_______________). Pull the bunched material out from underneath the victim and _______________ it around the victim.

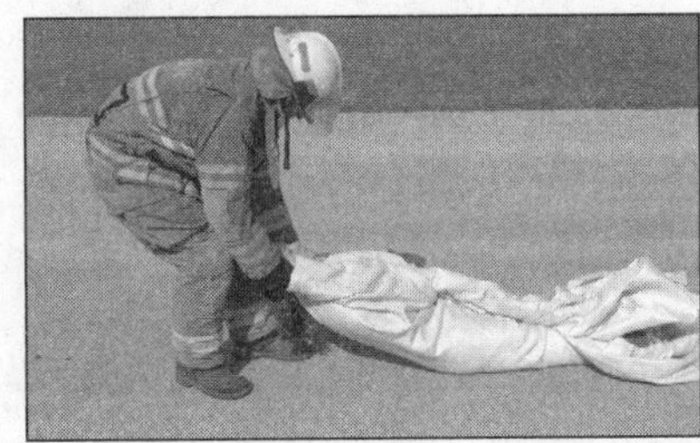

4. Grab the material at the _______________ and drag the victim _______________ to safety.

Skill Drill 12-9: Performing a Standing Drag

Test your knowledge of this skill drill by filling in the correct words in the photo captions.

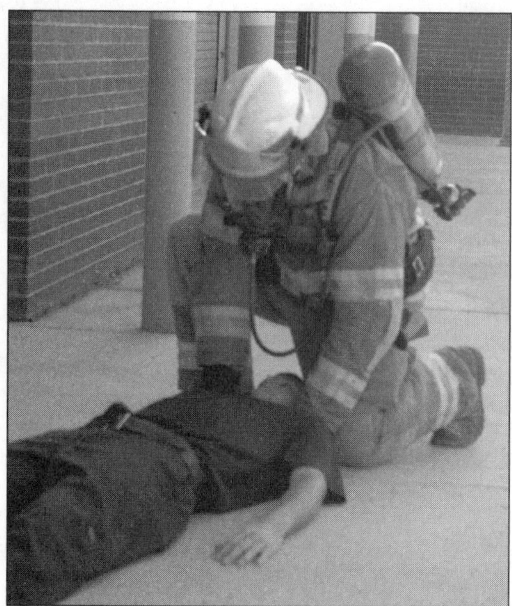

1. Kneel at the head of the _______________ victim.

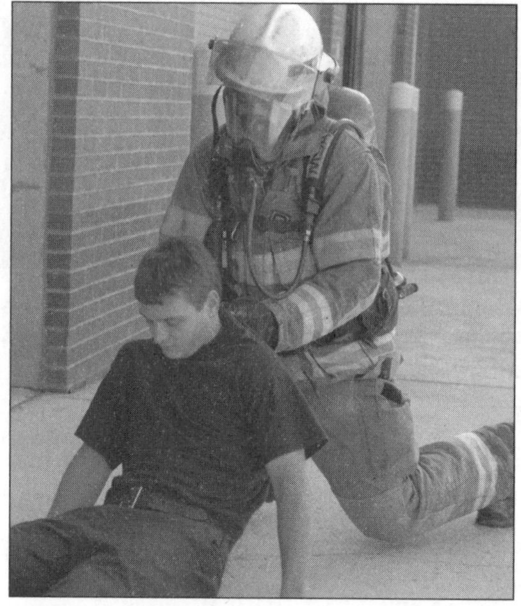

2. Raise the victim's head and torso by _______________ degrees, so that the victim is leaning against you.

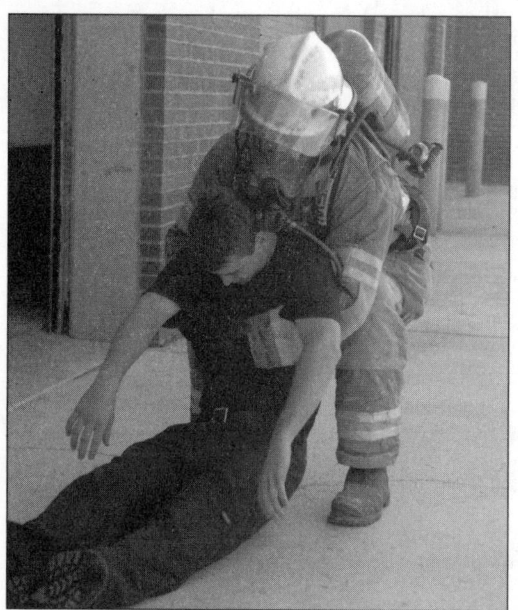

3. Reach under the victim's _______________, wrap your arms around the victim's _______________, and lock your arms.

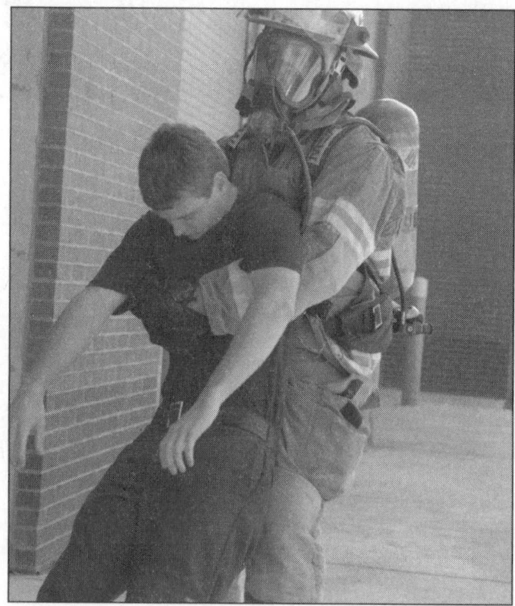

4. Stand straight up using your legs. _______________ the victim out.

Skill Drill 12-10: Performing a Webbing Sling Drag

Test your knowledge of this skill drill by filling in the correct words in the photo captions.

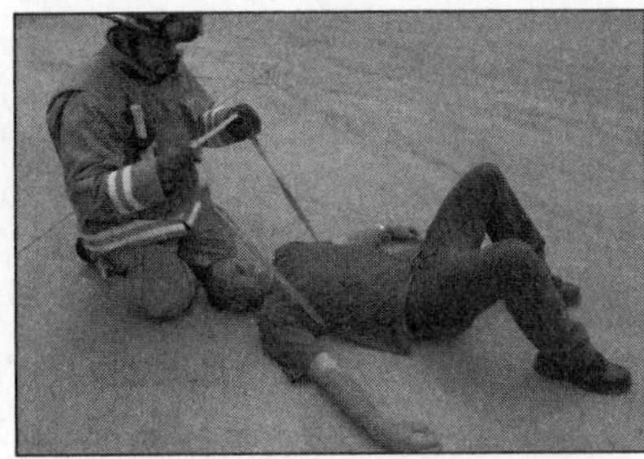

1. Place the victim in the _______________ of the loop so the webbing is behind the victim's _______________.

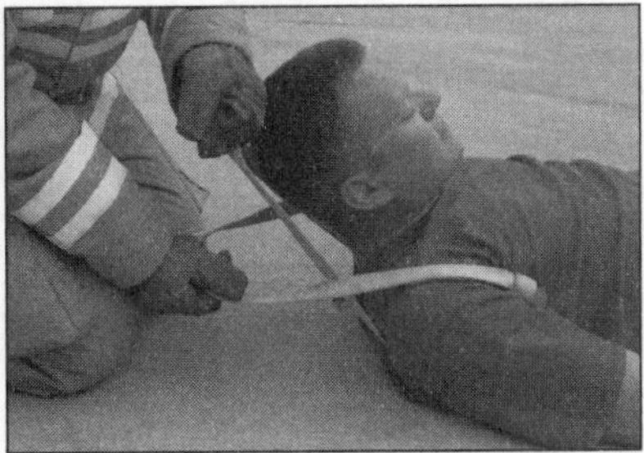

2. Take the _______________ loop over the victim and place it above the victim's head. Reach through, grab the webbing behind the victim's back, and pull through all the excess webbing. This creates a loop at the top of the victim's head and _______________ loops around the victim's arms.

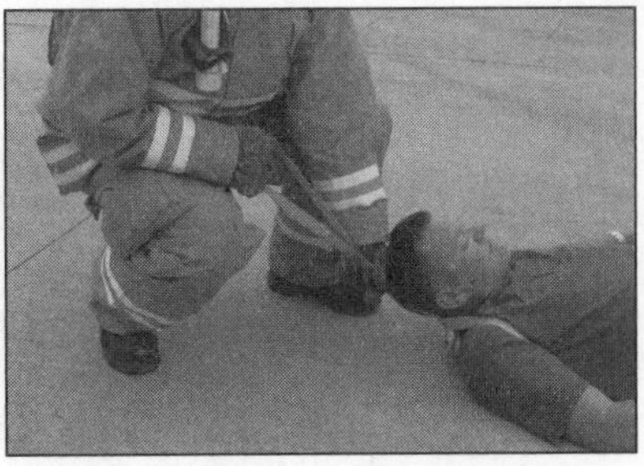

3. Adjust your hand placement to protect the victim's _______________ while dragging.

Skill Drill 12-11: Performing a Fire Fighter Drag

Test your knowledge of this skill drill by filling in the correct words in the photo captions.

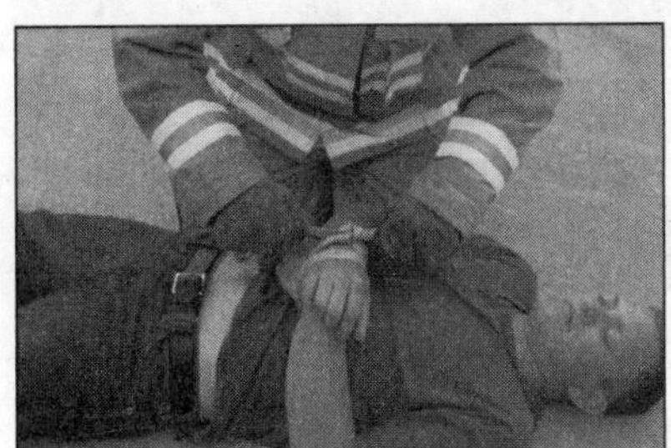

1. Tie the victim's wrists together with _______________ that is handy.

2. Get down on your hands and knees and _______________ the victim.

3. Pass the victim's tied hands around your neck, _______________ your arms, and drag the victim across the floor by crawling on your hands and knees.

Skill Drill 12-12: Performing a One-Person Emergency Drag from a Vehicle

Test your knowledge of this skill drill by filling in the correct words in the photo captions.

1. Grasp the victim under the arms and ______________ his or her head between your arms.

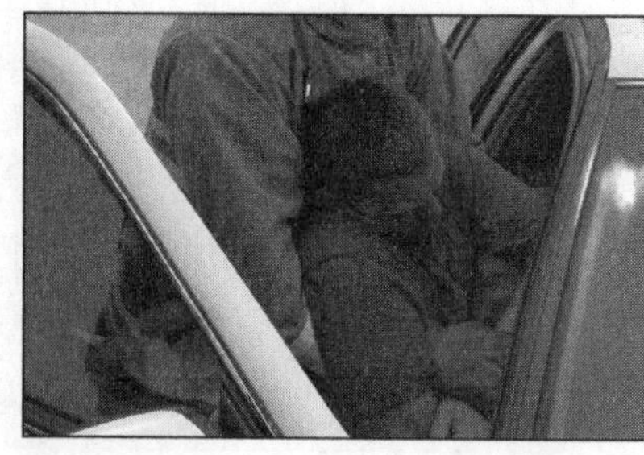

2. Gently ______________ the victim out of the vehicle.

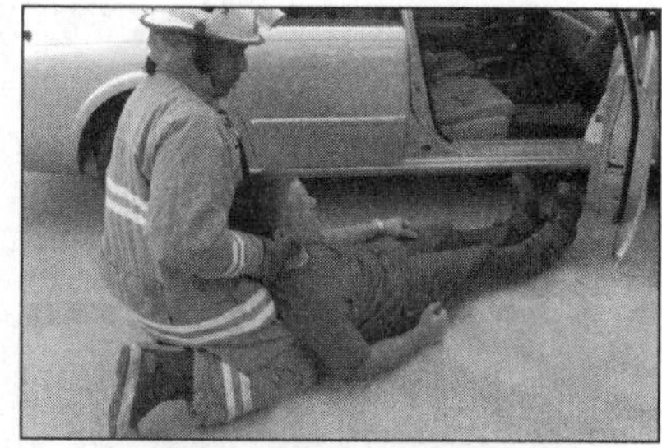

3. Lower the victim down into a ______________ position in a safe place.

Skill Drill 12-13: Performing a Long Backboard Rescue

Test your knowledge of this skill drill by placing the photos below in the correct order. Number the first step with a "1," the second step with a "2," and so on.

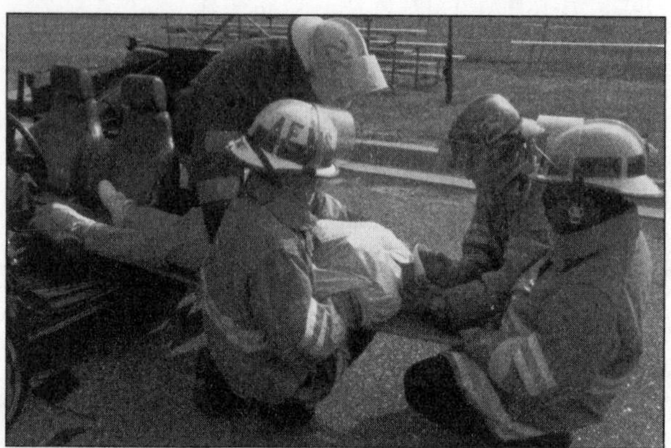

______ The third responder exits the vehicle and moves to the backboard opposite the second responder. Working together, they continue to slide the victim until the victim is fully on the backboard.

______ The second and third responders rotate the victim as a unit in several short, coordinated moves. The first responder (relieved by the fourth responder as needed) supports the victim's head and neck during rotation (and later steps).

______ The first responder provides in-line manual support of the victim's head and cervical spine.

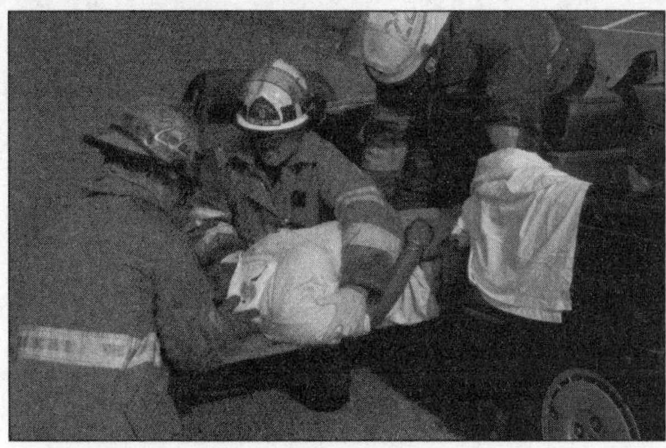

______ The third responder moves to an effective position for sliding the victim. The second and third responders slide the victim along the backboard in coordinated, 8" to 12" (20-cm to 31-cm) moves until the victim's hips rest on the backboard.

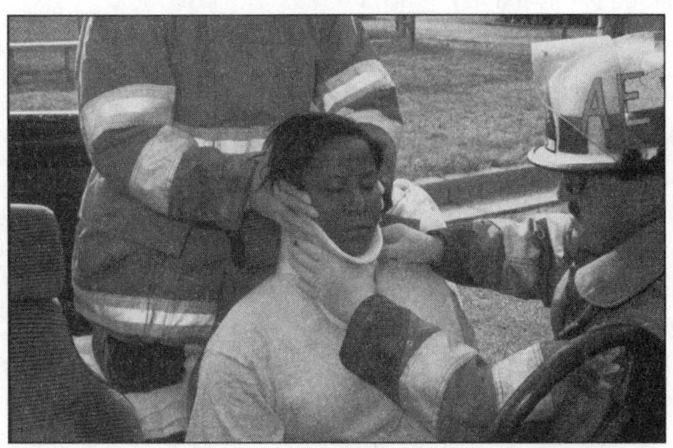

______ The second responder gives commands and applies a cervical collar.

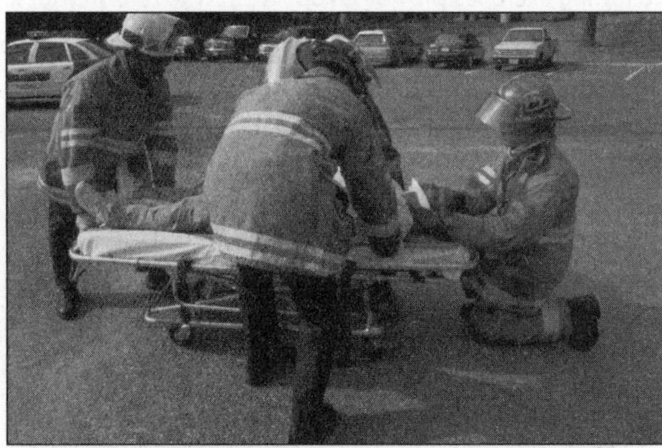

______ The first (or fourth) responder continues to stabilize the victim's head and neck, while the second, third, and fourth responders carry the victim away from the vehicle.

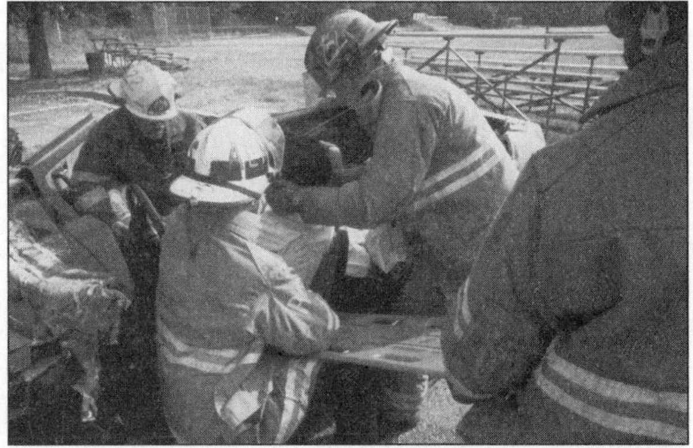

______ The first (or fourth) responder places the backboard on the seat against the victim's buttocks. The second and third responders lower the victim onto the long backboard.

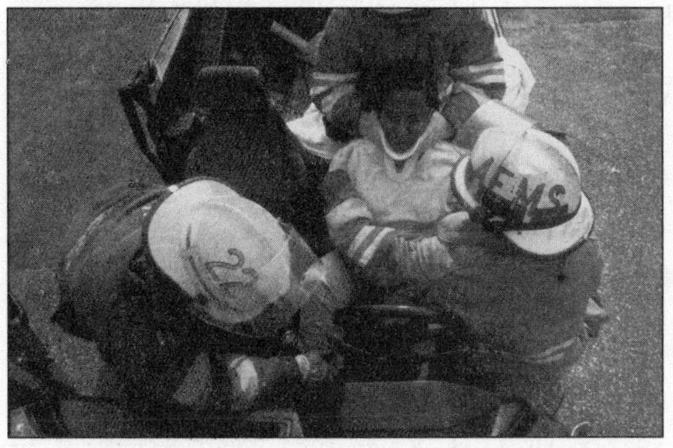

______ The third responder frees the victim's legs from the pedals and moves the legs together without moving the victim's pelvis or spine.

Mission-Specific Competencies: Response to Illicit Laboratories

Workbook Activities

The following activities have been designed to help you. Your instructor may require you to complete some or all of these activities as a regular part of your hazardous materials training program. You are encouraged to complete any activity that your instructor does not assign as a way to enhance your learning in the classroom.

Chapter Review

The following exercises provide an opportunity to refresh your knowledge of this chapter.

Matching

Match each of the terms in the left column to the appropriate definition in the right column.

______	**1.** Methamphetamine	**A.**	Bacterial agent
______	**2.** Anthrax	**B.**	Viral agent
______	**3.** Smallpox	**C.**	Ricin
______	**4.** Toxin	**D.**	Used to manufacture, process, culture, or synthesize an illegal drug
______	**5.** Illicit laboratory	**E.**	Also known as crank or ice
______	**6.** Sulfur mustard	**F.**	Blood agent
______	**7.** Cyanide	**G.**	Choking agent
______	**8.** Chlorine	**H.**	Blister agent
______	**9.** LEL	**I.**	Explosive material specialists
______	**10.** EOD	**J.**	May indicate a potentially flammable atmosphere

Multiple Choice

Read each item carefully, and then select the best response.

______ **1.** Which of the following is a potential chemical warfare agent?
- **A.** Anthrax
- **B.** Cyanide
- **C.** Ebola
- **D.** Ricin

______ **2.** Which of the following is a potential biological warfare agent?
- **A.** Sulfur mustard
- **B.** Anhydrous ammonia
- **C.** Botulinum
- **D.** Chlorine

______ **3.** All of the following are potential hazards of chemical warfare agents, except:
- **A.** difficulty breathing.
- **B.** nausea.
- **C.** dizziness.
- **D.** fever.

CHAPTER

13

______ 4. All of the following are potential hazards of biological warfare agents, except:

A. skin blisters or rash.

B. nausea.

C. nerve damage.

D. fever.

______ 5. All of the following describe forms of living microorganisms, except one of the following, which comprises by-products of living organisms.

A. Bacterial agents

B. Fungal agents

C. Viral agents

D. Toxins

Vocabulary

Define the following terms using the space provided.

1. Clandestine drug laboratory:

____

____

____

2. Explosive ordnance disposal (EOD) personnel:

____

____

____

3. Illicit laboratory:

____

____

____

Fill-in

Read each item carefully, and then complete the statement by filling in the missing word(s).

1. Red phosphorus labs and anhydrous ammonia labs are processes used to produce ______________.

2. Detonating cords may look like ______________ ______________ to an uninformed responder.

3. Biological agents typically have a greater time between entering the body and the onset of symptoms, possibly as long as ______________ ______________.

4. Depending on the quantity of drugs involved or the capacity of the illicit lab, the incident may be investigated by the federal ______________ ______________ ______________.

5. It is essential that whenever an explosive device is suspected to be involved, the appropriate ______________ ______________ ______________ personnel are notified.

True/False

If you believe the statement to be more true than false, write the letter "T" in the space provided. If you believe the statement to be more false than true, write the letter "F."

1. ______ A common characteristic of methamphetamine labs is the use of ephedrine and pseudoephedrine (cold medicine) tablets.
2. ______ Illicit laboratories can be small enough to fit inside the trunk of a car.
3. ______ Biological agents such as anthrax can be cultured illegally in illicit laboratories.
4. ______ Tablets are ground in household blenders as the first step in methamphetamine production.
5. ______ Clandestine drug laboratories will rarely have unusual chemical odors.
6. ______ Biological toxins are typically extracted from a plant or animal.
7. ______ Cyanide is typically extracted from the camphor bean, which gives it its characteristic smell.
8. ______ Personal protective equipment is routinely used by those operating clandestine drug laboratories.
9. ______ Canine teams used in law enforcement may need to be decontaminated.
10. ______ Decontamination areas and equipment should be established prior to any responder entering an illicit laboratory.

Short Answer

Complete this section with short written answers using the space provided.

1. List five methamphetamine chemicals and their legitimate uses.

____

____

____

____

____

Word Fun

The following crossword puzzle is an activity provided to reinforce correct spelling and understanding of terminology associated with hazardous materials. Use the clues provided to complete the puzzle.

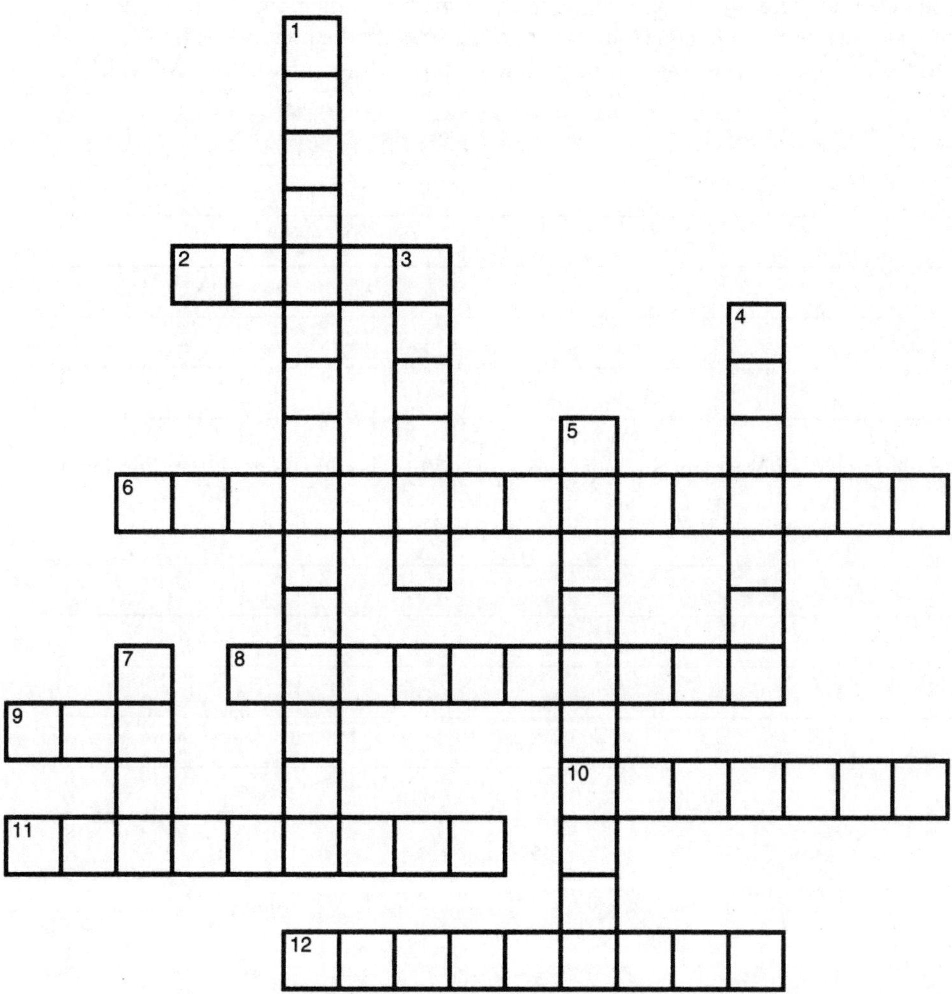

CLUES

Across

2 These have been used to provide perimeter defense around the outside of laboratories.

6 ________________ laboratories are illicit operations consisting of a sufficient combination of apparatus and chemicals that either has been or could be used in the manufacture or synthesis of controlled substances. (2 words)

8 ________________ are often used as mobile cooking laboratories. (2 words)

9 Responders must wear the appropriate ________________ while operating at illicit laboratories. (abbreviation)

10 A(n) ________________ laboratory is any unlicensed or illegal structure, vehicle, facility, or physical location that may be used to manufacture, process, culture, or synthesize an illegal drug, hazardous material/WMD device, or agent.

11 Because of the nature of criminal laboratories, it is always considered a prudent practice to include personnel properly trained in the recognition of ________________ devices as part of your hazard assessment plan.

12 Methamphetamine is a powerful ________________, and its users often suffer from extreme paranoia, sleeplessness, and anxiety.

Down

1 ________________ may look like rescue rope to an uninformed responder. (2 words)

3 A(n) ________________ basket may be used to carry those victims in a hazardous materials incident who cannot walk.

4 The ________________ of illicit laboratories are often partially or completely obscured.

5 ________________ warfare agents will generally be categorized in one of four ways: bacterial agents, fungal agents, viral agents, and toxins.

7 A mode of methamphetamine production that involves the use of red phosphorus, iodine, lye, and sulfuric acid. (2 words, abbreviation)

Fire Alarms

The following real case scenarios will give you an opportunity to explore the concerns associated with hazardous materials. Read each scenario, and then answer each question in detail.

1. You are the officer in charge of a fire company responding to a structure fire in an apartment building. On your arrival inside the premises, you note that there is paraphernalia such as terrorist training manuals, propaganda, and documents associated with a known terrorist organization. Your crew also notes the presence of timers, switches, fuses, gunpowder, and what appears to be rescue rope with wires protruding from the end. What is your next course of action?

____

____

____

____

____

____

2. You need to assist in decontaminating law enforcement S.W.A.T. personnel. What should you advise and do for them?

____

____

____

____

____

____

Skill Drills

Skill Drill 13-1: Identifying and/or Avoiding Potential Unique Safety Hazards

Test your knowledge of this skill drill by filling in the correct words below.

1. Visually assess the structure or property that is suspected to contain a ______________ operation for outward warning signs, such as the presence of security and ______________ systems (including triggering devices and booby traps), precursor ______________ containers, laboratory equipment, or hostile ______________.
2. Establish a safe ______________ ______________ based on the hazards identified.
3. Notify the appropriate ______________ ______________ personnel, technicians, and allied professionals based on the hazards identified.
4. Make an assessment of any ______________ who may be present and any ______________ they are presenting.

Skill Drill 13-2: Conducting Joint Hazardous Materials/EOD Operations

Test your knowledge of this skill drill by filling in the correct words below.

1. Discuss with law enforcement or EOD personnel those ______________ or ______________ that are potentially explosive and/or hazardous.
2. Develop a joint ______________ plan, if necessary, to render the device or materials safe for collection as ______________.
3. Develop a ______________ plan to support EOD personnel and equipment.

Mission-Specific Competencies: Air Monitoring and Sampling

Workbook Activities

The following activities have been designed to help you. Your instructor may require you to complete some or all of these activities as a regular part of your hazardous materials training program. You are encouraged to complete any activity that your instructor does not assign as a way to enhance your learning in the classroom.

Chapter Review

The following exercises provide an opportunity to refresh your knowledge of this chapter.

Matching

Match each of the terms in the left column to the appropriate definition in the right column.

_____	**1.** CGI	**A.**	A quick test carried out in the field to ensure the meter is operating correctly prior to entering a contaminated atmosphere
_____	**2.** Bump test	**B.**	Used to detect flammable and potentially explosive atmospheres
_____	**3.** Photo-ionization detectors	**C.**	The amount of oxygen in normal atmospheres
_____	**4.** 20.9%	**D.**	General survey instruments that detect vaporous chemicals at very low levels, even in the ppm range
_____	**5.** 19.5%	**E.**	Oxygen-enriched atmosphere
_____	**6.** 23.5%	**F.**	Oxygen-deficient atmosphere
_____	**7.** Radiation dosimeter	**G.**	Detects the presence of radiation
_____	**8.** Radiation detection device	**H.**	Accounts for the different types of gases that might be encountered
_____	**9.** CO detectors	**I.**	Records the potential dose of radiation the wearer might have received
_____	**10.** Relative response curve	**J.**	Measure the presence of carbon monoxide

Multiple Choice

Read each item carefully, and then select the best response.

_____ **1.** From the time an air sample is drawn into the machine, until the machine processes the sample and gives a reading, is referred to as the:

A. recovery time.
B. reaction time.
C. relative response factor.
D. calibration rate.

_____ **2.** Which of the following is often referred to as "sewer gas?"

A. Sulfur mustard
B. Hydrogen sulfide
C. Hydrogen
D. Hydrogen cyanide

_____ **3.** Which of the following quickly deadens the sense of smell to the point where an individual may no longer be able to detect the gas by its odor?
A. Sulfur mustard
B. Hydrogen sulfide
C. Hydrogen
D. Hydrogen cyanide

_____ **4.** Which of the following is appropriate for measuring flammable atmospheres at or below their LEL/LFL in air?
A. PID
B. GC
C. CGI
D. FID

_____ **5.** The IDLH exposure limit for carbon monoxide is:
A. 35 ppm.
B. 10,000 ppm.
C. between 12% and 75%.
D. 1200 ppm.

Labeling

Label the following diagrams with the correct terms.

1. Types of detectors and monitors.

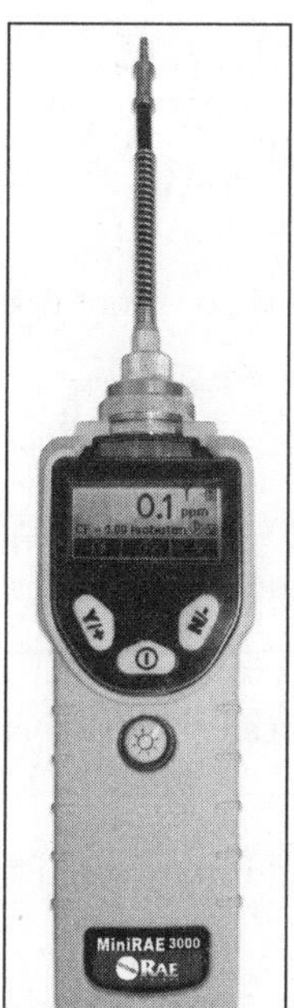

A. ____________________

B. ____________________

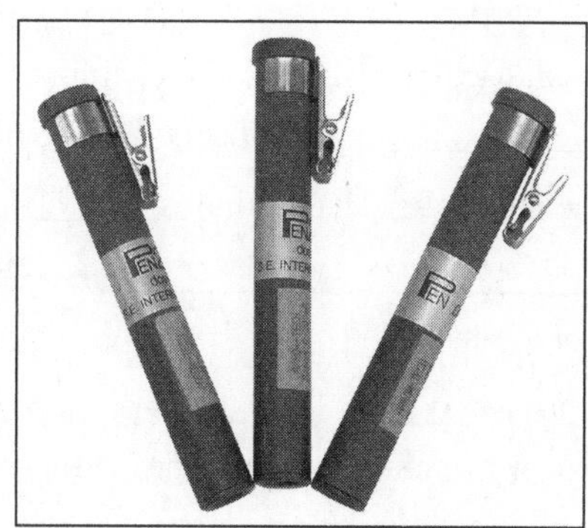

C. ____________________

Vocabulary

Define the following terms using the space provided.

1. Volatile organic compound:

____

____

____

2. Relative response factor:

____

____

____

3. Situational awareness (SA):

____

____

____

Fill-in

Read each item carefully, and then complete the statement by filling in the missing word(s).

1. _______________ is the term used to describe the process of ensuring that a particular instrument will respond appropriately to a predetermined concentration of gas.
2. _______________ _______________ is a term that describes focusing on observing and understanding the visual clues available, orienting oneself to those inputs relative to the current situation, and making rapid decisions based on those inputs.
3. The manufacturer tests the monitor against various gases and vapors and provides a _______________ _______________ _______________ that can be used to determine the correct percentage of the gas being monitored for.
4. Ambient air at sea level contains _______________ % oxygen.
5. _______________ _______________ is a chemical paper that allows the user to determine if a liquid or vapor is an acid or a base.
6. A device is "_______________" when it begins its operational period in a clean atmosphere by displaying normal values (or no values) or when the device recovers to that same baseline state after exposure to a gas or vapor.
7. The _______________ _______________ of a particular device is a function of how much time it takes a detector or monitor to clear so a new reading can be taken.
8. In atmospheres where the CO concentration is in the vicinity of _______________ ppm, all emergency personnel should be wearing SCBA.
9. _______________ _______________ _______________ look somewhat like pH paper but can perform several tests at once.
10. _______________-_______________ _______________ are capable of detecting several different hazards at the same time.

True/False

If you believe the statement to be more true than false, write the letter "T" in the space provided. If you believe the statement to be more false than true, write the letter "F."

1. ______ The MSA CGI was the first and most popular monitoring instrument used in the fire service.
2. ______ Too little oxygen creates a health risk, but too much oxygen creates an elevated fire risk.
3. ______ The highest level of safety exists when the atmosphere reaches 100% of the LEL/LFL.
4. ______ Colorimetric tubes are designed to detect single substances and/or chemical families or groups.
5. ______ A carbon monoxide sensor may pick up the presence of hydrogen sulfide, hydrogen, or hydrogen cyanide.
6. ______ The pH scale goes from 0 (strong base) on the low end of the scale, to 14 (strong acid) on the opposing end of the scale.
7. ______ Electrochemical sensors have a shelf life of 5 to 7 years, depending on the use.
8. ______ In some cases, radiation detection devices can identify which type of radiation is present (alpha, beta, gamma).
9. ______ Personal dosimeters stay on the responder throughout an incident.
10. ______ Reaction time for a device can be as short as 1 or 2 seconds or as long as 60 to 90 seconds.

Short Answer

Complete this section with short written answers using the space provided.

1. List the 10 basic actions for detection and monitoring as identified in NFPA 472.

____

____

____

____

____

____

____

____

____

____

CLUES

Word Fun

The following crossword puzzle is an activity provided to reinforce correct spelling and understanding of terminology associated with hazardous materials. Use the clues provided to complete the puzzle.

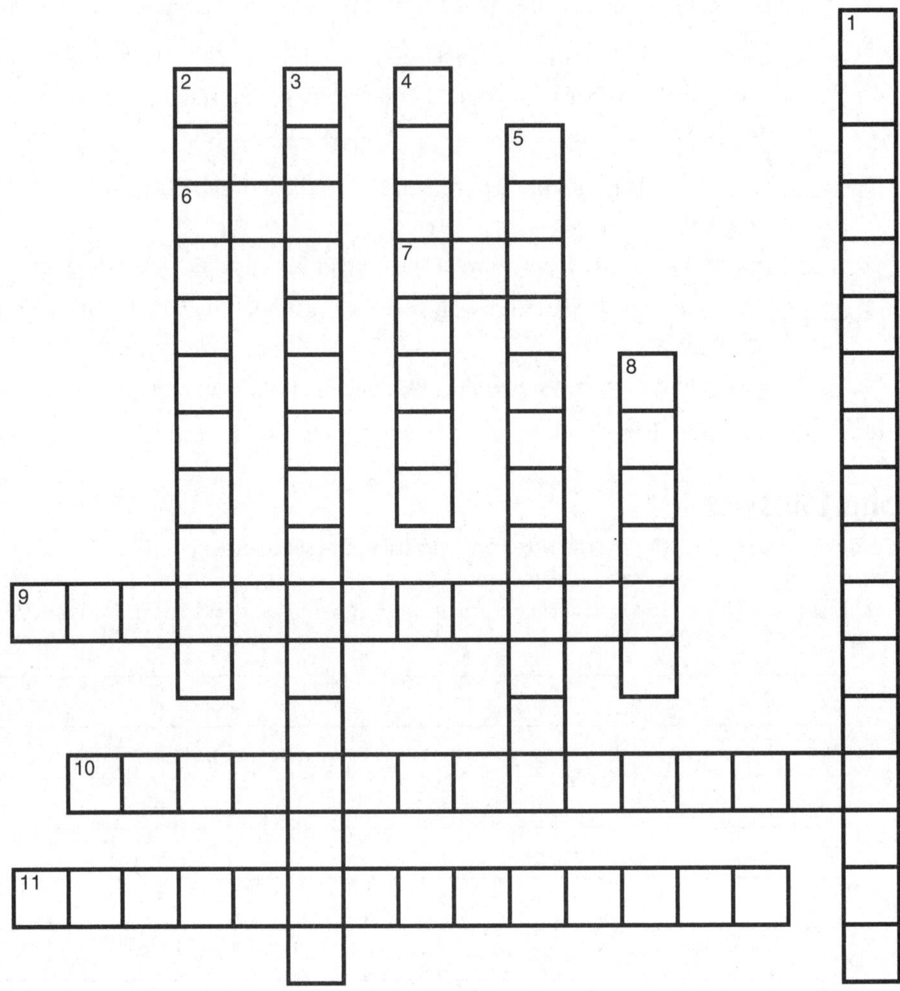

Across

6 The minimum amount of gaseous fuel that must be present in the air for the air/fuel mixture to be flammable or explosive. (abbreviation)

7 An instrument that provides real-time measurements of airborne contaminant levels of volatile organic compounds. (abbreviation)

9 The amount of time it takes a detector/monitor to clear itself so a new reading can be taken.

10 A(n) ______________ monitor is a single-gas device using a specific toxic gas sensor to detect and measure levels of H_2S in an airborne environment. (2 words)

11 The first and most popular detection instrument used in the fire service was the MSA2a ______________ indicator. (2 words)

Down

1 A device that measures the amount of radioactive exposure incurred by an individual. (2 words)

2 The process of ensuring that a particular detection/monitoring instrument will respond appropriately to a predetermined concentration of gas. An accurately calibrated machine ensures that the device is detecting the gas or vapor it is intended to detect, *at a given level*.

3 The ______________ factor is a mathematical computation that must be completed for a detector to correlate the differences between the gas that is used to calibrate the device and the gas that is being detected in the atmosphere.

4 A quick field test to ensure a detection/monitoring meter is operating correctly prior to entering a contaminated atmosphere.

5 pH is an expression of the amount of dissolved ______________ (H^+) in a solution. (2 words)

8 A(n) ______________ basket may be used to carry those victims in a hazardous materials incident who cannot walk.

Fire Alarms

The following real case scenarios will give you an opportunity to explore the concerns associated with hazardous materials. Read each scenario, and then answer each question in detail.

1. You've responded to a spill in a chemistry laboratory at a local college. On arrival, you are alerted that there are 5 one-gallon glass bottles that have broken and have spilled on the floor. The graduate student who reported the incident states that the material involved is tetrahydrofuran. He had attempted to clean it up himself for approximately 30 minutes, but stopped after experiencing irritated eyes, headache, nausea, and dizziness. The MSDS indicates a flashpoint of 6°F (–14°C), an LEL of 2%, and a UEL of 11.8%. Your personnel indicate that they have measured 10% on their combustible gas indicator and 300 parts per million on direct reading instruments. In order to evacuate the building, security personnel pulled a manual fire alarm, which automatically shuts down the ventilation system as well. Outdoor weather conditions are approximately 95°F (35°C) and humid. What potential explosion risk would you determine based on the above scenario?

2. The MSDS indicates a NIOSH REL of 200 ppm and a short-term exposure limit of 250 ppm. Based on your direct reading instrument, how does the graduate student's exposure to airborne concentrations compare with the NIOSH exposure limits?

Skill Drills

Skill Drill 14-3: Using a Multi-Gas Meter to Provide Atmospheric Monitoring (after the proper level of PPE is selected)

Test your knowledge of this skill drill by filling in the correct words in the photo captions.

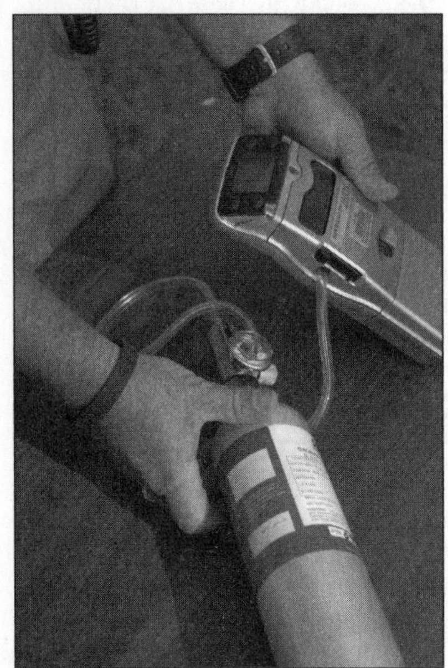

1. Understand the manufacturer's recommendations and local standard operating procedures for ______________-______________ ______________ use. Turn on the device and zero it in a clean atmospheric environment. Let the device warm up. Perform a ______________ ______________.

2. Approach the hazardous material and ______________ the atmosphere.

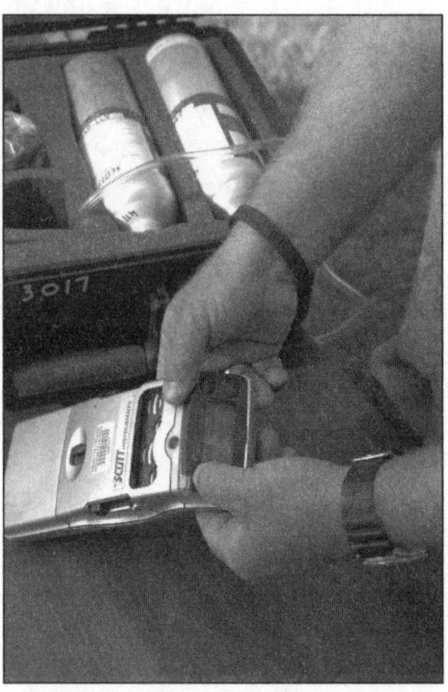

3. Interpret the meaning of the ______________. Return to a safe atmosphere. Return the meter to zero and follow the appropriate procedures to turn the meter off and return the meter to ______________.

Answer Key

Chapter 1: Hazardous Materials: Overview

Matching

1. F (page 6)
2. A (page 10)
3. G (page 5)
4. C (pages 5, 15)
5. B (page 5)
6. E (page 5)
7. D (page 12)
8. J (page 11)
9. H (page 11)
10. I (page 11)

Multiple Choice

1. B (page 5)
2. B (page 5)
3. D (page 6)
4. D (page 11)
5. C (page 11)
6. A (page 11)
7. C (page 11)
8. A (page 6)
9. B (page 9)
10. C (page 10)

Vocabulary

1. **Local emergency planning committee (LEPC):** A group consisting of members of industry, transportation officials, the public at large, media, and fire and police agencies that gathers and disseminates information on hazardous materials stored in the community and ensures that there are adequate local resources to respond to a chemical event in the community. (page 11)
2. **Material safety data sheet (MSDS):** A form, provided by manufacturers and compounders (blenders) of chemicals, containing information about chemical composition, physical and chemical properties, health and safety hazards, emergency response, and waste disposal of the material. (page 11)
3. **Specialist level:** A level of fire fighter expertise at which the individual receives more specialized training than does a hazardous materials technician. Practically speaking, the two levels are not very different. Most of the training that specialist employees receive is either product- or transportation-mode specific. (page 10)
4. **HAZWOPER:** Hazardous Waste Operations and Emergency Response. This OSHA regulation governs hazardous materials waste sites and response training. Specifics can be found in 29 CFR 1910.12. Subsection (q) is specific to emergency response. (page 6)
5. **Operations level:** The level at which the responder should be able to recognize a potential hazardous materials incident, isolate the incident, deny entry to other responders and the public, and take defensive actions such as shutting off valves and protecting drains without having contact with the product. Operations-level responders act in a defensive fashion. (page 8)

Fill-in

1. Environmental Protection Agency (pages 5, 15)
2. material safety data sheet (page 11)
3. more (page 1)
4. Preplanning (page 13)
5. protective (page 8)
6. technicians (page 10)
7. NFPA 472 (page 6)
8. HAZWOPER (page 6)
9. Hazardous waste (page 5)
10. state-plan (page 6)

True/False

1. T (page 6)
2. F (page 11)
3. T (page 11)
4. T (page 12)
5. F (page 12)
6. T (page 12)
7. F (page 13)
8. T (page 12)
9. F (pages 5, 15)
10. T (page 11)

Short Answer

1. *Awareness level:* Persons who, in the course of their normal duties, could encounter an emergency involving hazardous materials/weapons of mass destruction (WMD). They are expected to recognize the presence of the hazardous materials/WMD, protect themselves, call for trained personnel, and secure the area. This level of training enables those who are first on scene of an incident to recognize a potential hazardous materials emergency, isolate the area, and call for assistance. Awareness-level-trained persons are not seen as responders, but they do take protective actions.

   Operations level: Persons who respond to hazardous materials/WMD incidents for the purpose of protecting nearby persons, the environment, or property from the effects of the release. Fire fighters in modern society are usually trained to the operations level because they should be able to recognize potential hazardous materials incidents, isolate and deny entry to other responders and the public, evacuate persons in danger, and take defensive actions such as shutting off valves and protecting drains without having contact with the product. Operations-level responders act in a defensive fashion.

   Technician level: Fire fighters who are trained to enter heavily contaminated areas using the highest levels of personal protection. Hazardous materials technicians take offensive actions.

   Specialist level: Fire fighters who receive more specialized training than do hazardous materials technicians. Practically speaking, however, the two levels are not very different. The majority of the specialized training relates to a specific product such as chlorine or to a specific mode of transportation such as rail emergencies.

   Hazardous materials officer level: A level of training intended for those assuming command of a hazardous materials incident beyond the operations level. Individuals trained as hazardous materials officers receive operations-level training as well as additional training specific to commanding a hazardous materials incident. The hazardous materials officer is trained to act as a branch director or group supervisor for the hazardous materials component of the incident. (pages 6–10)

2. The Superfund Amendments and Reauthorization Act (SARA) was one of the first laws to affect how fire departments respond in a hazardous materials emergency. Finalized in 1986, SARA was the original driver for OSHA's HAZWOPER regulation. Indirectly, it indicated that workers handling hazardous wastes should have a minimum amount of training. Additionally, this law laid the foundation that ultimately allowed local fire departments and the community at large to obtain information on how and where hazardous materials were stored in their community. (pages 11, 15)

Word Fun

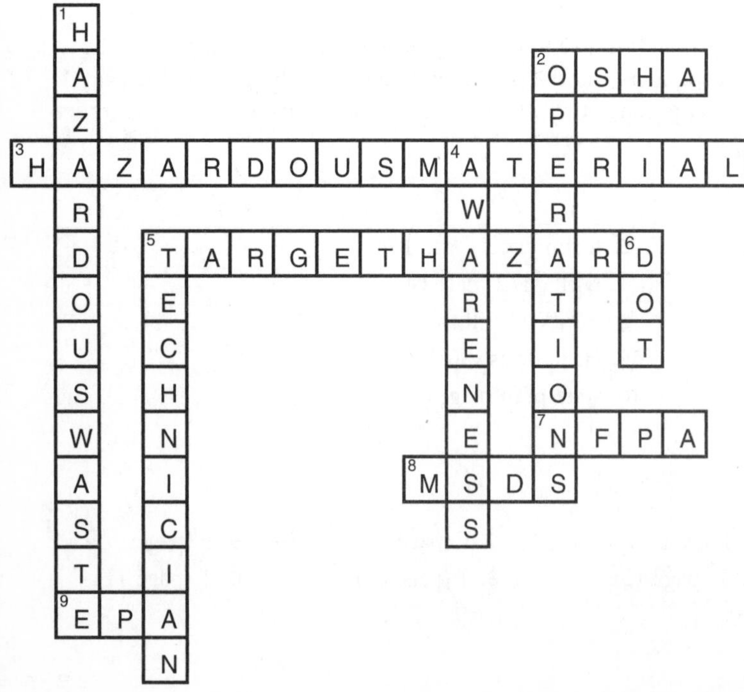

Fire Alarms

1. A hazardous material, as defined by the U.S. Department of Transportation (DOT), is a material that poses an unreasonable risk to the health and safety of operating emergency personnel, the public, and/or the environment if it is not properly controlled during handling, storage, manufacture, processing, packaging, use and disposal, or transportation. (page 5)
2. The response begins with finding out about potential hazards in your area. Response agencies also should conduct incident planning activities related to target hazards and other potential problem areas throughout the jurisdiction. Preplanning activities enable agencies to develop logical and appropriate response procedures for anticipated incidents. Jurisdictions that have no railways or maritime ports do not have to include training for those kinds of responses. Planning should focus on the real threats that exist in your community or adjacent communities you could be assisting. Once the threats have been identified, agencies must determine how they will respond. Some agencies establish parameters that guide their response to particular hazardous material incidents. Those parameters outline incident severity based on the nature of the chemical, the amount released, or the type of occupancy involved in the incident. (pages 12, 13)

Chapter 2: Hazardous Materials: Properties and Effects

Matching

1. E (page 23)
2. F (page 25)
3. A (page 28)
4. H (page 38)
5. B (page 25)
6. D (page 27)
7. C (page 30)
8. I (page 25)
9. G (page 37)
10. J (page 23)

Multiple Choice

1. A (page 22)
2. C (page 22)
3. D (page 23)
4. C (page 23)
5. B (page 23)
6. C (page 26)
7. A (page 27)
8. B (page 27)
9. A (page 26)
10. C (page 28)
11. A (page 29)
12. D (page 29)
13. B (page 29)
14. C (page 29)
15. A (page 30)
16. A (page 31)
17. D (page 32)
18. B (page 35)
19. C (page 35)
20. B (page 36)
21. B (page 38)
22. A (page 38)

Labeling

1. Vapor density

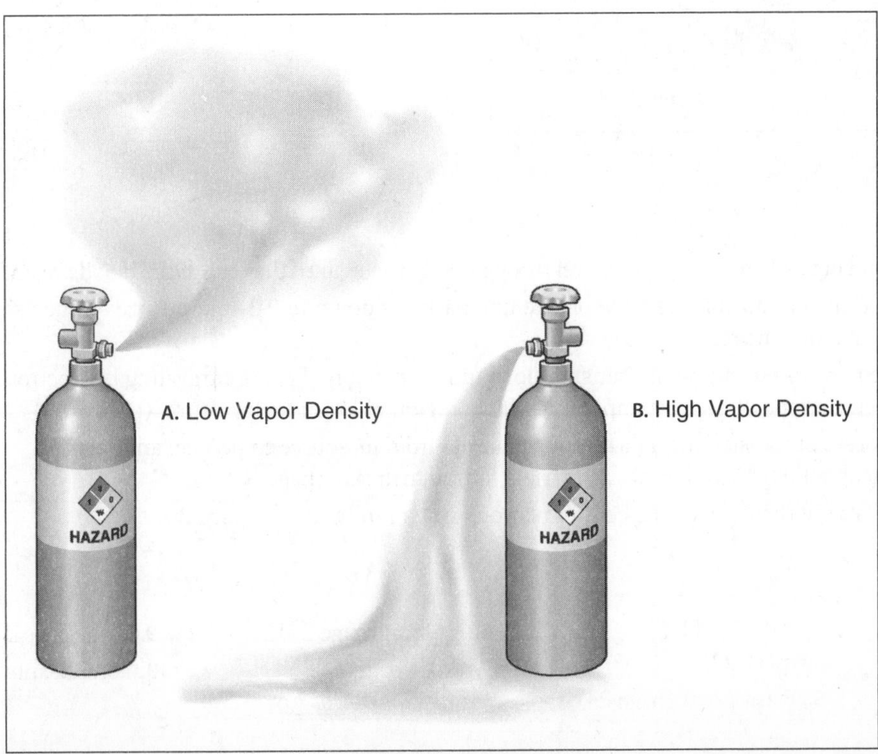

2. Radiation

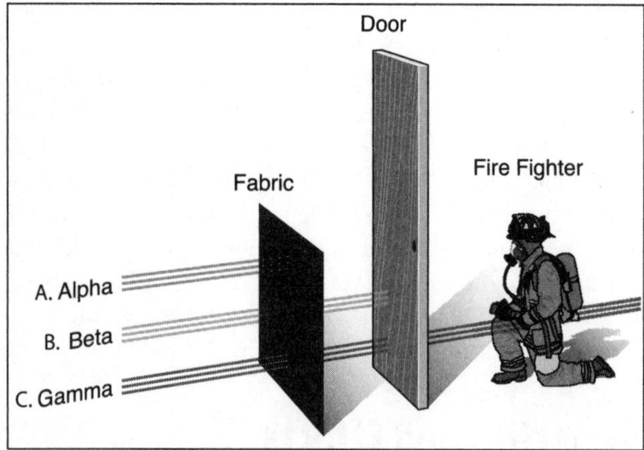

3. Four ways a chemical substance can enter the body

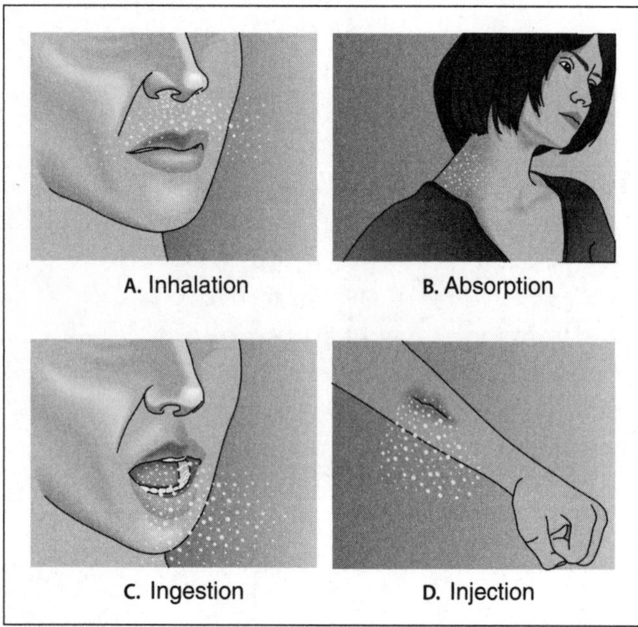

Vocabulary

1. **Absorption:** The exposure route where substances travel through body tissue until they reach the bloodstream. (page 37)
2. **HEPA:** High-efficiency particulate air filter capable of catching particles down to 0.3-micron size—much smaller than a typical dust or alpha radiation particle. (page 31)
3. **Radiation:** The combined process of emission, transmission, and absorption of energy traveling by electromagnetic wave propagation between a region of higher temperature and a region of lower temperature. (page 30)
4. **Contamination:** The process of transferring a hazardous material from its source to people, animals, the environment, or equipment, all of which may act as carriers for the material. (page 32)
5. **Nerve agent:** A toxic substance that attacks the central nervous system in humans. (page 34)

Fill-in

1. predict (page 22)
2. physical (page 23)
3. 14.7 (page 26)
4. –45°F (–43°C) (page 24)
5. Flash point (page 24)
6. ignition (page 25)
7. float (page 27)
8. elements (page 30)
9. Beta (page 31)
10. convulsants (page 35)

True/False

1. T (page 23)
2. F (page 26)
3. T (page 24)
4. T (page 25)
5. T (page 26)
6. F (pages 30–32)
7. F (page 30)
8. T (page 41)
9. T (page 34)
10. T (page 38)

Short Answer

1. *Thermal:* Heat created from intentional explosions or fires, or cold generated by cryogenic liquids.

   Radiological: Radioactive contamination from dirty bombs; alpha, beta, and gamma radiation.

   Asphyxiation: Oxygen deprivation caused by materials such as nitrogen; tissue asphyxiation from blood agents.

   Chemical: Injury and death caused by the intentional release of toxic industrial chemicals, nerve agents, vesicants, poisons, or other chemicals.

   Etiological: Illness and death resulting form biohazards such as anthrax, plague, and smallpox; hazards posed by bloodborne pathogens.

   Mechanical: Property damage and injury caused by explosion, falling debris, shrapnel, firearms, explosives, and slips, trips, and falls.

   Psychogenic: The mental harm from being potentially exposed to, contaminated by, and even just being in close proximity to an incident of this nature. (pages 32, 34)
2. S: Salivation
   L: Lachrymation (tearing)
   U: Urination
   D: Defecation
   G: Gastric disturbance
   E: Emesis (vomiting)
   M: Miosis (constriction of the pupil) (page 34)
3. (1) *Inhalation:* Through the lungs. (2) *Absorption:* By permeating the skin. (3) *Ingestion:* Via the gastrointestinal tract. (4) *Injection:* Through cuts or other breaches in the skin. (pages 35, 36)
4. (1) The amount of radiation absorbed by the body has a direct relationship to the degree of damage done. (2) The amount of exposure time ultimately determines the extent of the injury. (page 32)
5. H: Hydrogen
   H: Helium
   H: Hydrogen cyanide
   H: Hydrogen fluoride
   M: Methane
   E: Ethylene
   D: Diborane
   I: Illuminating gas (methane/ethane mixture)
   C: Carbon monoxide
   A: Ammonia
   N: Neon
   N: Nitrogen
   A: Acetylene (page 27)

Word Fun

		1 V												
		A		2 S				3 B			4 G		5 I	
6 S		P		7 A	L	P	H	A			A		N	
E		O		R				S			M		J	
8 C	A	R	C	I	N	O	G	E	9 N		M		E	
O				N					E		A		C	
N							10 B		R		R		T	
D		11 N					E		V		12 A	C	I	D
A		E					T		E		D		O	
R		U		13 I	N	H	A	L	A	T	I	O	N	
Y		T							G		A			
		R				14 I	N	G	E	S	T	I	O	N
		O							N		I			
15 C	O	N	T	A	M	I	N	A	T	I	O	N		
		S							S		N			

Fire Alarms

1. Make every effort to reduce or eliminate the ability of the substance to enter your body, and keep the duration of the exposure to an absolute minimum. This requires you to reduce the time you are exposed to the material, to stay far enough away so that you are not directly exposed, and/or to shield yourself with personal protective equipment or solid objects. Time, distance, and shielding are methods used to protect fire fighters from the adverse effects of radiation. If you suspect a radiation incident at a fixed facility, you should initially ask for the radiation safety officer of the facility. This person is responsible for maintaining the use, handling, and storage procedures for all of the radioactive material at the site. This person will likely be a tremendous resource to you and will know exactly what is being used at the facility. (pages 30–32)
2. From a terrorism perspective, irritants may be employed to incapacitate rescuers or to drive a group of people into another area where a more dangerous substance can be released. Irritants pose the least amount of danger in terms of toxicity of all the potential WMD agents a fire fighter may encounter. Exposed patients can be decontaminated with clean water, and the residual effects of the exposure should not be significant. (pages 32–35)

Chapter 3: Recognizing and Identifying the Hazards

Matching

1. H (page 67)
2. F (page 70)
3. G (page 55)
4. C (page 51)
5. J (page 55)
6. A (page 70)
7. E (page 54)
8. B (page 54)
9. I (page 59)
10. D (page 53)

Multiple Choice

1. D (page 52)
2. D (page 52)
3. B (page 53)
4. D (page 54)
5. A (page 54)
6. C (page 55)
7. A (page 55)
8. C (page 55)
9. D (page 56)
10. C (page 56)
11. B (page 58)
12. A (page 58)
13. C (page 60)
14. D (pages 61, 62)
15. A (pages 61, 62)
16. B (page 70)
17. A (page 70)
18. C (page 73)
19. A (page 74)
20. B (page 75)

Labeling

1. Chemical transport vehicles.

A. An MC-306 flammable liquid tanker (page 56)

B. An MC-307 chemical hauler (page 57)

C. An MC-312 corrosives tanker (page 57)

D. An MC-331 pressure cargo tanker (page 57)

E. An MC-338 cryogenic tanker (page 57)

F. A tube trailer (page 58)

G. A dry bulk cargo tank (page 58)

H. A nonpressurized rail tank car (page 58)

Vocabulary

1. **Shipping papers:** A shipping order, bill of lading, manifest, or other shipping document serving a similar purpose and usually including the names and addresses of both the shipper and the receiver as well as a list of shipped materials along with their quantity and weight. (page 67)
2. **Secondary containment:** Any device or structure that prevents environmental contamination when the primary container or its appurtenances fail. Examples of secondary containment include dikes, curbing, and double-walled tanks. (page 52)

3. **Hazardous materials:** Any materials or substances that pose an unreasonable risk of damage or injury to persons, property, or the environment if not properly controlled during handling, storage, manufacture, processing, packaging, use and disposal, or transportation. (page 50)
4. **Pipeline right of way:** An area, patch, or roadway that extends a certain number of feet on either side of the pipe itself and that may contain warning and informational signs about hazardous materials carried in the pipeline. (page 59)
5. **Placards and labels:** Placards are diamond-shaped indicators (10" [25 cm] on each side) that must be placed on all four sides of highway transport vehicles, railroad tank cars, and other forms of transportation carrying hazardous materials. Labels are smaller versions (4" [10-cm] diamond-shaped indicators) of placards and are used on the four sides of individual boxes and smaller packages being transported. Placards and labels are intended to give fire fighters a general idea of the hazard inside a particular container. A placard may identify the broad hazard class (e.g., flammable, poison, corrosive) of material that a tanker contains, while the label on a box inside a delivery truck relates only to the potential hazard inside that package. (page 60)

Fill-in

1. size-up (page 50)
2. 102 (page 54)
3. 312 (page 57)
4. tube (page 58)
5. pipelines (page 59)
6. ~~W~~ (page 61)
7. Hazardous Materials Information System (HMIS) (page 63)
8. material safety data sheet (MSDS) (page 67)
9. military (page 63)
10. waybills, consist (page 70)

True/False

1. T (page 50)
2. T (page 53)
3. F (pages 53–54)
4. T (page 54)
5. F (page 59)
6. F (page 62)
7. T (page 62)
8. T (page 50)
9. T (page 67)
10. F (page 70)

Short Answer

1. The following items are included on a pesticide bag label:
   - Name of the product
   - Active ingredients
   - Hazard statement
   - Total amount of product in the container
   - Manufacturer's name and address
   - U.S. Environmental Protection Agency (EPA) registration number, which provides proof that the product was registered with the EPA
   - The EPA establishment number, which shows where the product was manufactured
   - Signal words to indicate the relative toxicity of the material:
     Danger—Poison: Highly toxic by all routes of entry
     Danger: Severe eye damage or skin irritation
     Warning: Moderately toxic
     Caution: Minor toxicity and minor eye damage or skin irritation
   - Practical first-aid treatment description
   - Directions for use
   - Agricultural use requirements
   - Precautionary statements such as mixing directions or potential environmental hazards
   - Storage and disposal information
   - Classification statement on who may use the product

   In addition, every pesticide label must carry the statement, "Keep out of reach of children." (page 55)
2. (1) *DOT Class 1:* Explosives. (2) *DOT Class 2:* Gases. (3) *DOT Class 3:* Flammable combustible liquids. (4) *DOT Class 4:* Flammable solids. (5) *DOT Class 5:* Oxidizers. (6) *DOT Class 6:* Poisons (including blood agents and choking agents). (7) *DOT Class 7:* Radioactive materials. (8) *DOT Class 8:* Corrosives. (9) *DOT Class 9:* Other regulated materials. (page 66)
3. *Yellow section:* Chemicals in this section are listed numerically by their four-digit UN identification number. Entry number 1017, for example, identifies chlorine. Use the yellow section when the UN number is known or can be identified. The entries include the name of the chemical and the emergency action guide number.

   Blue section: Chemicals in the blue section are listed alphabetically by name. The entry will include the emergency action guide number and the identification number. The same information, organized differently, appears in both the blue and yellow sections.

Orange section: This section contains the emergency action guides. Guide numbers are organized by general hazard class and indicate what basic emergency actions should be taken, based on hazard class.

Green section: This section is organized numerically by UN identification number and provides the initial isolation distances for specific materials. Chemicals included in this section are highlighted in the blue or yellow sections. Any materials listed in the green section are always extremely hazardous. This section also directs the reader to consult the tables listing toxic inhalation hazard materials (TIH). These gases or volatile liquids are extremely toxic to humans and pose a hazard to health during transportation. (page 64)

4. The NFPA 704 hazard identification system uses a diamond-shaped symbol of any size, which is itself broken into four smaller diamonds, each representing a particular property or characteristic. The blue diamond at the nine o'clock position indicates the health hazard posed by the material. The top red diamond indicates flammability. The yellow diamond at the three o'clock position indicates reactivity. The bottom white diamond is used for special symbols and handling instructions. The blue, red, and yellow diamonds will each contain a numerical rating of 0 to 4, with 0 being the least hazardous and 4 being the most hazardous for that type of hazard. The white quadrant will not have a number but may contain special symbols. Among the symbols used are a burning O (oxidizing capability), a three-bladed fan (radioactivity), and a W (W̶) with a slash through it (water reactive). (page 61)
5. (1) Physical and chemical characteristics; (2) physical hazards of the material; (3) health hazards of the material; (4) signs and symptoms of exposure; (5) routes of entry; (6) permissible exposure limits; (7) responsible-party contact; (8) precautions for safe handling (including hygiene practices, protective measures, and procedures for cleaning up spills or leaks); (9) applicable control measures, including personal protective equipment; (10) emergency and first-aid procedures; (11) appropriate waste disposal (page 67)

Word Fun

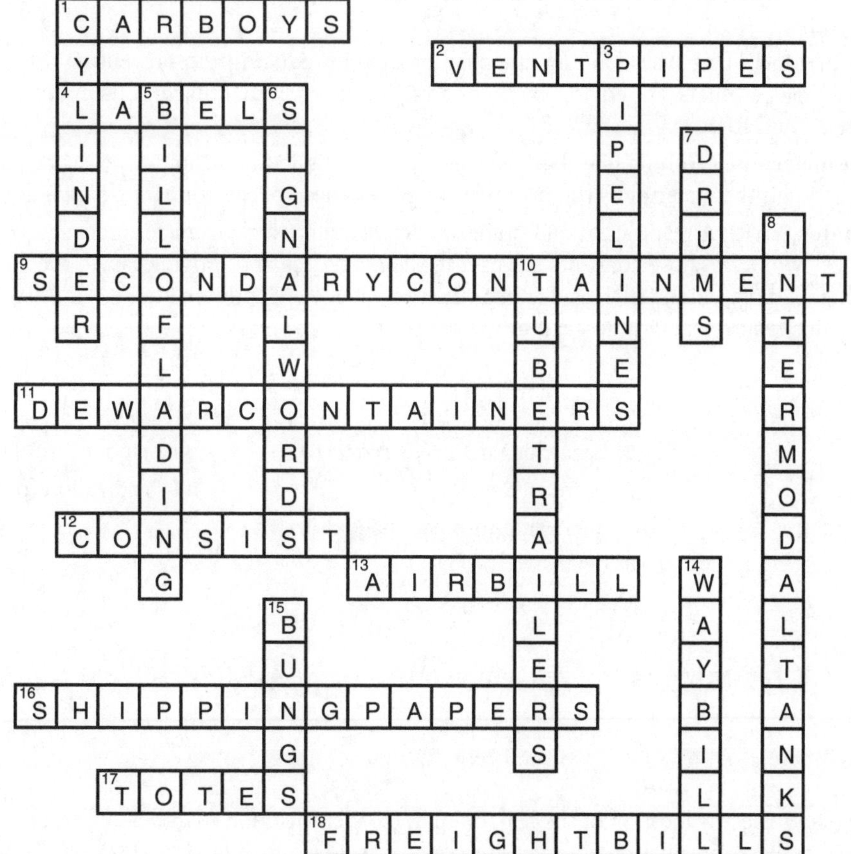

Fire Alarms

1. (a) Name of the caller and callback telephone number; location of the actual incident or problem; name of the chemical involved in the incident (if known); shipper or manufacturer of the chemical (if known); container type; railcar or vehicle markings or numbers; shipping carrier's name; recipient of material; local conditions and exact description of the situation

   (b) 1-800-424-9300 (page 70)

Chapter 4: Estimating Potential Harm and Planning a Response

Matching

1. B (page 85)
2. A (page 92)
3. C (page 85)
4. D (pages 85, 86)
5. E (page 85)

Multiple Choice

1. D (page 90)
2. B (page 90)
3. A (page 91)
4. A (page 92)
5. D (page 99)
6. A (page 91)
7. D (page 91)
8. B (page 90)
9. C (page 101)
10. D (page 92)

Labeling

1. A (page 96)
2. D (page 98)
3. C (page 98)
4. B (page 97)

Vocabulary

1. **Defensive objectives:** Actions that do not involve the actual stopping of the leak, release of a hazardous material, or contact of responders with the material. These actions include preventing further injury and controlling or containing the spread of the hazardous material. (page 92)
2. **Isolation of the hazard area:** Steps taken to identify a perimeter around a contaminated atmosphere. Isolating an area is driven largely by the nature of the released chemicals and the environmental conditions that exist at the time of the release. (page 88)
3. **Decontamination corridor:** The physical and/or chemical process of reducing and preventing the spread of contaminants from people, animals, the environment, or equipment involved at hazardous materials/weapons of mass destruction incidents. (page 101)
4. **Chemical-resistant materials:** Clothing (suit-fabrics) specifically designed to inhibit or resist the passage of chemicals into and through the material by the processes of penetration, permeation, or degradation. (page 95)
5. **Supplied-air respirator (SAR):** A respirator that obtains its air through a hose from a remote source such as a compressor or storage cylinder. A hose connects the user to the air source and provides air to the face piece. SARs are useful during extended operations such as decontamination, clean-up, and remedial work. Also referred to as positive-pressure air-line respirators (with escape units). (page 100)

Fill-in

1. resources (page 84)
2. Alkaline (page 90)
3. characteristics (page 84)
4. Skin absorption (page 90)
5. hazardous materials (page 101)
6. Litmus (page 90)
7. Sheltering-in-place (page 89)
8. defensive (page 92)
9. responders (page 91)
10. Permeation (page 95)

True/False

1. F (page 95)
2. T (page 86)
3. T (page 86)
4. F (page 97)
5. F (page 100)

Short Answer

1. (1) The exact address and specific location of the leak or spill; (2) identification of indicators and markers of hazardous materials; (3) all color or class information obtained from placards; (4) four-digit United Nations/North American Hazardous Materials Code numbers for the hazardous materials; (5) hazardous material identification obtained from shipping papers or MSDS and the potential quantities of hazardous materials involved; (6) description of the container, including its size, capacity, type, and shape; (7) the amount of chemical that could leak and the amount that has already leaked; (8) exposures of people and the presence of special populations (children or elderly); (9) the environment in the immediate area; (10) current weather conditions, including wind direction and speed; (11) a contact or callback telephone number and two-way radio frequency or channel. (page 90)

2. (1) Isolate the area affected by the leak or spill, and evacuate victims who could become exposed to the hazardous material if the leak or spill were to progress. (2) Control where the spill or release is spreading. (3) Contain the spill to a specific area. (page 92)
3. The special technical group may include a second safety officer. This hazardous materials safety officer is responsible for the hazardous materials team's safety only. The group may also include a hot zone entry team, a decontamination team, a backup entry team (rapid intervention team), and a hazardous materials information research team. (page 124)
4. *Safe atmosphere:* No harmful hazardous materials effects exist, so personnel can handle routine emergencies without donning specialized PPE.

   Unsafe atmosphere: A hazardous material that is no longer contained has created an unsafe condition or atmosphere. A person who is exposed to the material for long enough will probably experience some form of acute or chronic injury.

   Dangerous atmosphere: Serious, irreversible injury or death may occur in the environment. (page 86)
5. *Level A protection:* Personal protective equipment provides protection against vapors, gases, mists, and even dusts. The highest level of protection, it requires a totally encapsulating suit that includes self-contained breathing apparatus.

   Level B protection: Personal protective equipment is used when the type and atmospheric concentration of substances have been identified. It generally requires a high level of respiratory protection but less skin protection (chemical-protective coveralls and clothing, chemical protection for shoes, gloves, and self-contained breathing apparatus outside of a nonencapsulating chemical-protective suit).

   Level C protection: Personal protective equipment is used when the type of airborne substance is known, the concentration is measured, the criteria for using air-purifying respirators are met, and skin and eye exposure is unlikely. It consists of standard work clothing with the addition of chemical-protective clothing, chemically resistant gloves, and a form of respirator protection.

   Level D protection: Personal protective equipment is used when the atmosphere contains no known hazard, and work functions preclude splashes, immersion, or the potential for unexpected inhalation of or contact with hazardous levels of chemicals. It is primarily a work uniform that includes coveralls and affords minimal protection. (pages 96–98)

Word Fun

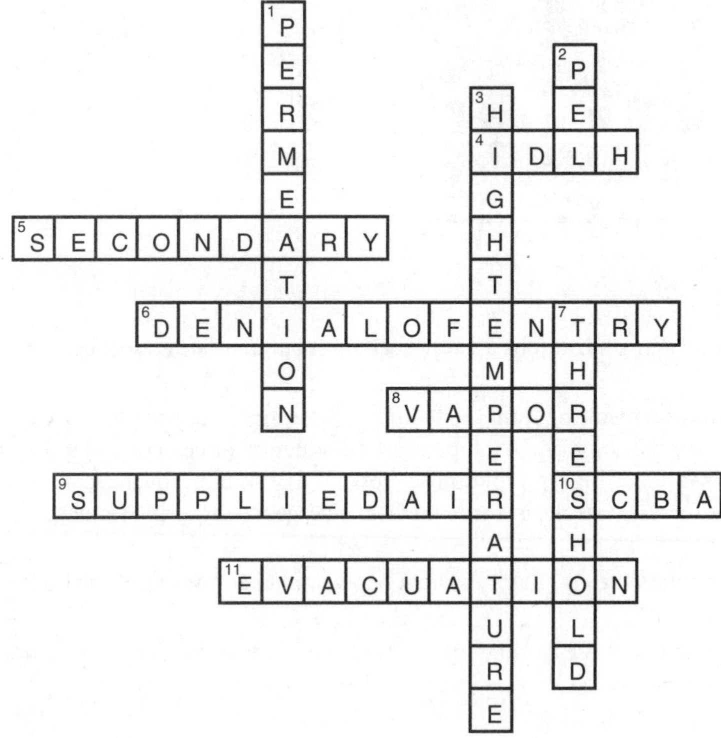

Fire Alarms

1. Contact the engineer; he or she may have some information and the waybill that you need. Once you have gathered your on-scene information, call CHEMTREC and the National Response Center.

2. The diamond marking will give you a general idea of the flammability, toxicity, reactivity, and any special hazards associated with the materials that are stored in the chemical storage shed. Collect the information and limit access to the building until the hazardous materials team arrives.

Chapter 5: Implementing the Planned Response

Matching

1. C (page 114)
2. D (page 116)
3. F (page 123)
4. G (page 114)
5. J (page 122)
6. A (page 114)
7. E (page 123)
8. B (page 124)
9. I (page 111)
10. H (page 124)

Multiple Choice

1. B (page 112)
2. D (page 117)
3. A (page 116)
4. C (page 117)
5. B (page 117)
6. C (page 117)
7. B (page 122)
8. A (page 119)
9. B (page 120)
10. C (page 120)
11. B (page 112)
12. A (page 112)
13. B (page 114)
14. C (page 114)
15. D (page 114)

Labeling

1. Control zones.

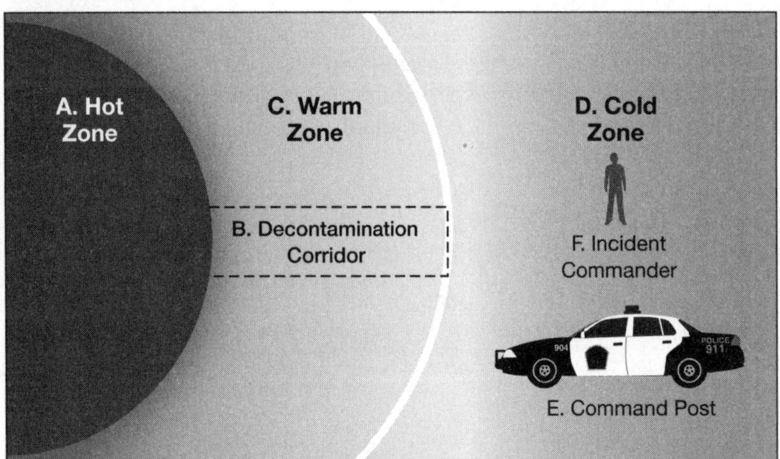

Vocabulary

1. **Shelter-in-place:** A method of safeguarding people in a hazardous area by keeping them in a safe atmosphere, usually inside structures. (page 117)
2. **Heat stroke:** A severe and sometimes fatal condition resulting from the failure of the temperature-regulating capacity of the body. It is caused by prolonged exposure to the sun or high temperatures. Reduction or cessation of sweating is an early symptom; body temperature of 105°F (41°C) or higher, rapid pulse, hot and dry skin, headache, confusion, unconsciousness, and convulsions may also occur. Heat stroke is a true medical emergency requiring immediate transport to a medical facility. (page 119)
3. **Heat exhaustion:** A mild form of shock caused when the circulatory system begins to fail as a result of the body's inadequate effort to give off excessive heat. (page 119)
4. **Backup personnel:** Individuals who remove or rescue those working in the hot zone in the event of an emergency. (page 116)

Fill-in

1. life (page 115)
2. safe area (page 116)
3. toxicity (page 116)
4. safety officer (page 123)
5. buddy system (page 116)
6. defensive (page 111)
7. 240 (page 119)
8. socks (page 120)
9. Size-up (page 111)
10. incident commander (page 114)

True/False

1. T (page 117)
2. T (page 114)
3. T (page 115)
4. F (page 112)
5. F (page 116)
6. T (page 124)
7. T (page 114)
8. F (page 118)
9. T (page 114)
10. T (page 114)

Short Answer

1. Some of the key benefits of using ICS are common terminology, consistent organization structure, consistent position titles, and common incident facilities. (page 120)
2. *Operations:* The Operations Section carries out the objectives developed by the IC and is responsible for all tactical operations at the incident.

   Planning: The Planning Section is responsible for the collection, evaluation, dissemination, and use of information relevant to the incident.

   Logistics: The Logistics Section can be viewed as the support side of an incident management structure and is responsible for providing facilities, services, and materials for the incident.

   Finance: The Finance/Administration Section tracks the costs related to the incident, handles procurement issues, records the time that responders are on the incident for billing purposes, and keeps a running cost of the incident. (pages 123–126)
3. *Hot zone:* The area immediately surrounding a hazardous materials spill/incident site that is directly dangerous to life and health. All personnel working in the hot zone must wear complete, appropriate protective clothing and equipment. Entry requires approval by the IC or a designated hazardous materials officer. Complete backup, rescue, and decontamination teams must be in place at the perimeter before operations begin.

   Warm zone: The area located between the hot zone and the cold zone at the incident. Personal protective equipment is required when working in this zone. The decontamination corridor is located in the warm zone, which is also called the contamination reduction zone.

   Cold zone: A safe area at a hazardous materials incident for those agencies involved in the operations. The incident commander, incident command post, EMS providers, and other support functions necessary to control the incident should be located in the cold zone, which is also called the clean zone or support zone. (pages 114–115)

Word Fun

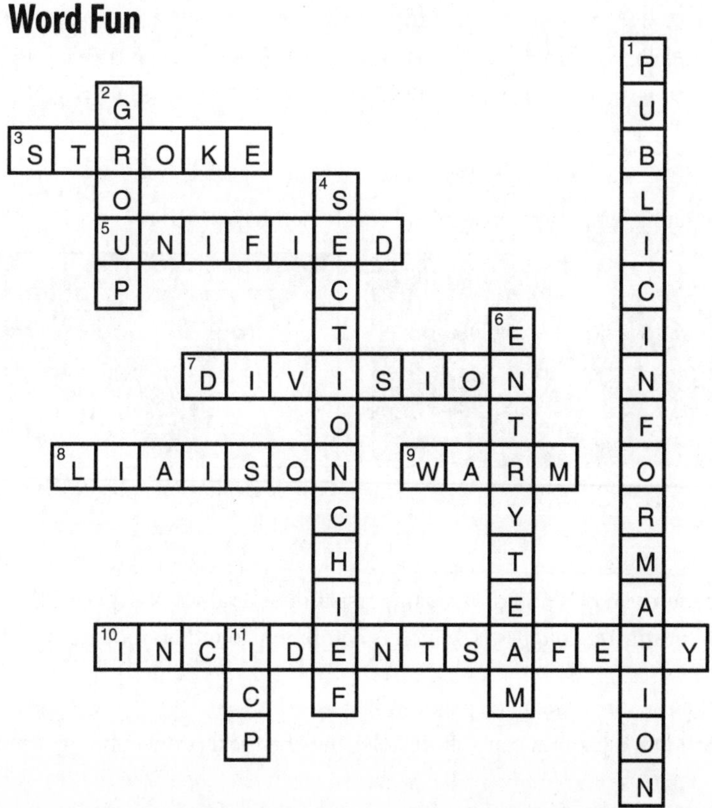

Fire Alarms

1. (a) A secondary device is a second explosive device placed in a location with the intent to kill emergency responders after an initial explosion has taken place.

   (b) Indicators of potential secondary devices may include trip devices such as timers, wires, or switches. They may also include common concealment containers, such as briefcases, backpacks, boxes, or other common packages, and uncommon concealment containers, such as pressurized vessels (e.g., propane tanks) or industrial chemical containers (e.g., chlorine storage containers). Attackers may be watching the site of the primary devices and waiting to manually activate the secondary devices. (page 75)

Skill Drills

Skill Drill 5-1: Performing an Emergency Decontamination

1. Ensure that you have the appropriate PPE. Stay clear of the product, and do not make physical contact with it. Make an effort to contain **runoff**. Instruct or assist victims in removing contaminated clothing.
2. Rinse **victims** with copious amounts of water. Avoid using water that is too warm or too cold; room-temperature water is best.

Chapter 6: Terrorism

Matching

1. E (page 145)
2. M (page 139)
3. A (page 143)
4. B (page 150)
5. K (page 138)
6. N (page 144)
7. O (page 142)
8. J (page 148)
9. G (page 142)
10. F (page 145)
11. C (page 141)
12. L (page 141)
13. D (pages 151, 154)
14. H (page 146)
15. I (page 141)

Multiple Choice

1. D (page 141)
2. A (page 137)
3. B (page 138)
4. C (page 137)
5. A (page 138)
6. C (page 139)
7. B (page 139)
8. D (page 141)
9. B (page 138)
10. C (page 146)
11. B (page 148)
12. C (page 148)
13. D (page 149)
14. A (page 150)
15. C (page 151)

Labeling

1. Based on the symptoms shown in these photographs, what type of agent was each victim exposed to?

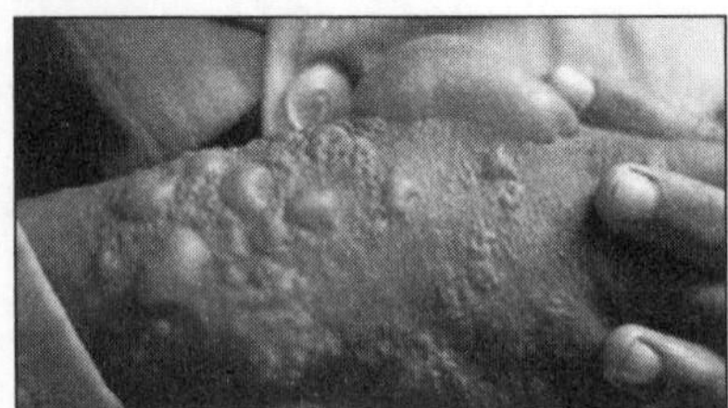

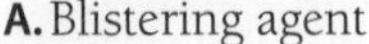

A. Blistering agent

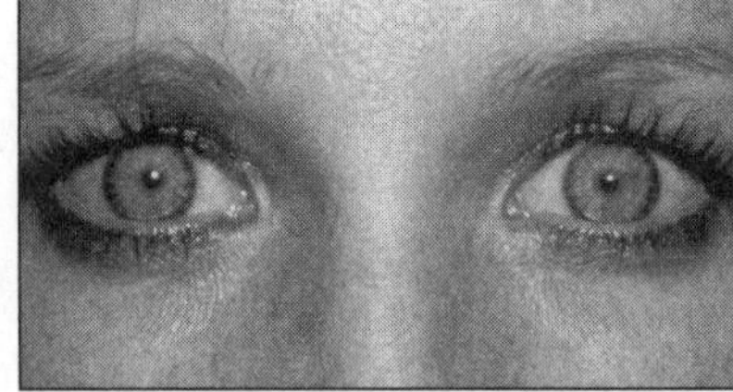

B. Nerve agent

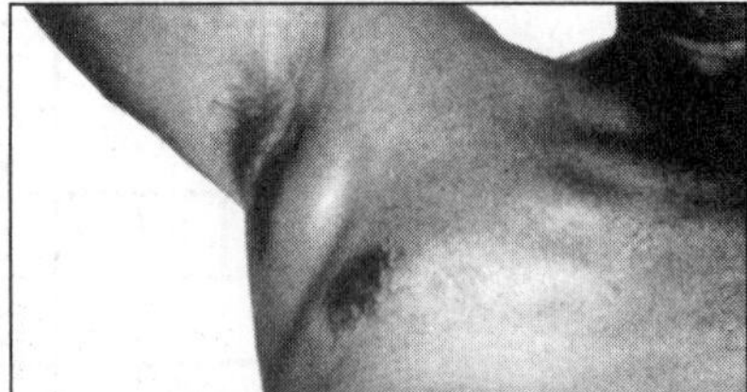

C. Biological agent (plague)

Vocabulary

1. **V-agent:** A nerve agent, principally a contact hazard; an oily liquid that can persist for several weeks. (page 143)
2. **Plague:** An infectious disease caused by the bacterium *Yersinia pestis,* which is commonly found on rodents. (page 146)
3. **Smallpox:** A highly infectious disease caused by the virus *Variola*. (page 147)
4. **Tabun:** A nerve gas that is both a contact hazard and a vapor hazard. It disables the chemical connection between the nerves and their target organs. (page 143)
5. **Universal precautions:** Procedures for infection control that treat blood and certain bodily fluids as capable of transmitting bloodborne diseases. (page 147)

6. **Forward staging area:** A strategically placed area, close to the incident site, where personnel and equipment can be held in readiness for rapid response to an emergency event. (page 139)
7. **Radiation dispersal device:** Any device that causes the purposeful dissemination of radioactive material without a nuclear detonation; a dirty bomb. (page 149)

Fill-in

1. Terrorism (page 134)
2. pipe bomb (page 138)
3. Nerve agents (page 142)
4. SLUDGE (page 143)
5. Crop-dusting (page 142)
6. incubation period (page 146)
7. Radioactive materials (page 148)
8. absorption (page 150)
9. mass decontamination (page 150)
10. perimeter (page 150)

True/False

1. F (page 146)
2. T (page 150)
3. F (page 148)
4. F (page 148)
5. T (page 148)
6. F (page 143)
7. F (page 138)
8. T (page 135)
9. T (page 149)
10. T (page 148)

Short Answer

1. Terrorists are motivated by a cause and choose targets they believe will help them achieve their goals and objectives. Terrorist incidents aim to instill fear and panic among the general population and to disrupt daily ways of life. Given this goal, terrorists tend to choose symbolic targets, such as places of worship, embassies, monuments, or prominent government buildings. Sometimes the objective is sabotage—that is, to destroy or disable a facility that is significant to the terrorist cause. The ultimate goal could be to cause economic turmoil by interfering with transportation, trade, or commerce. (page 135)
2. Ecoterrorism refers to illegal acts committed by groups supporting environmental or related causes. Examples include spiking trees to sabotage logging operations, vandalizing a university research laboratory that is conducting experiments on animals, or firebombing a store that sells fur coats.

   Groups of terrorists could engage in cyberterrorism by electronically attacking government or private computer systems. This type of terrorism would disrupt many day-to-day activities in our society because the use of computers is woven into most things we do as part of contemporary life.

   Agroterrorism includes the use of chemical or biological agents to attack the agricultural industry or the food supply. The deliberate introduction of an animal disease such as foot-and-mouth disease to the livestock population could result in major losses to the food industry and produce fear among members of the general population. (pages 137–138)
3. Your first priority should be to ensure the safety of the scene. During the initial stages of an incident, you will not know whether the event was caused by an intentional act or by accidental circumstances. In any incident involving an explosion, follow departmental procedures to ensure the safety of rescuers, victims, and bystanders. Consider the possibility that a secondary device may be in the vicinity. Quickly survey the area for any suspicious bags, packages, or other items.

   It is also possible that chemical, biological, or radiological agents may be involved in a terrorist bombing. Qualified personnel with monitoring instruments should be assigned to check the area for potential contaminants. The initial size-up should also include an assessment of hazards and dangerous situations. The stability of any building involved in the explosion must be evaluated before anyone is permitted to enter it. Entering an unstable area without proper training and equipment may complicate rescue and recovery efforts. (page 139)
4. Responding to a terrorist incident puts fire fighters and other emergency personnel at risk. Although responders must ensure their own safety at every incident, a terrorist incident may carry an extra dimension of risk. Because the terrorist's objective is to cause as much harm as possible, emergency responders are just as likely to be targets as are ordinary civilians.

   In most cases, the first emergency units will not be dispatched for a known WMD or terrorist incident. Rather, the initial dispatch might be for an explosion, for a possible hazardous materials incident, for a single person with difficulty breathing, or for multiple victims with similar symptoms. Emergency responders will usually not know that a terrorist incident has occurred until personnel on the scene begin to piece together information gained from their own observations and from witnesses. (page 149)
5. There are three ways to limit exposure to radioactivity: Keep the time of the exposure as short as possible; stay as far away (distance) from the source of the radiation as possible; use shielding to limit the amount of radiation absorbed by the body. Alpha particles quickly lose their energy and, therefore, can travel only 1" to 2" (3-cm to 5-cm) from their source. Clothing or a sheet of paper can stop this type of energy. If ingested or inhaled, alpha particles can damage a number of internal organs.

Beta particles are more powerful, capable of traveling 10' to 15' (3-m to 5-m). Heavier materials such as metal, plastic, and glass can stop this type of energy. Beta radiation can harm both the skin and the eyes, and its ingestion or inhalation will damage internal organs.

Gamma rays can travel significant distances, penetrate most materials, and pass through the body. This type of radiation is the most destructive to the human body. The only materials with sufficient mass to stop gamma radiation are concrete, earth, and dense metals such as lead. (page 148)

Word Fun

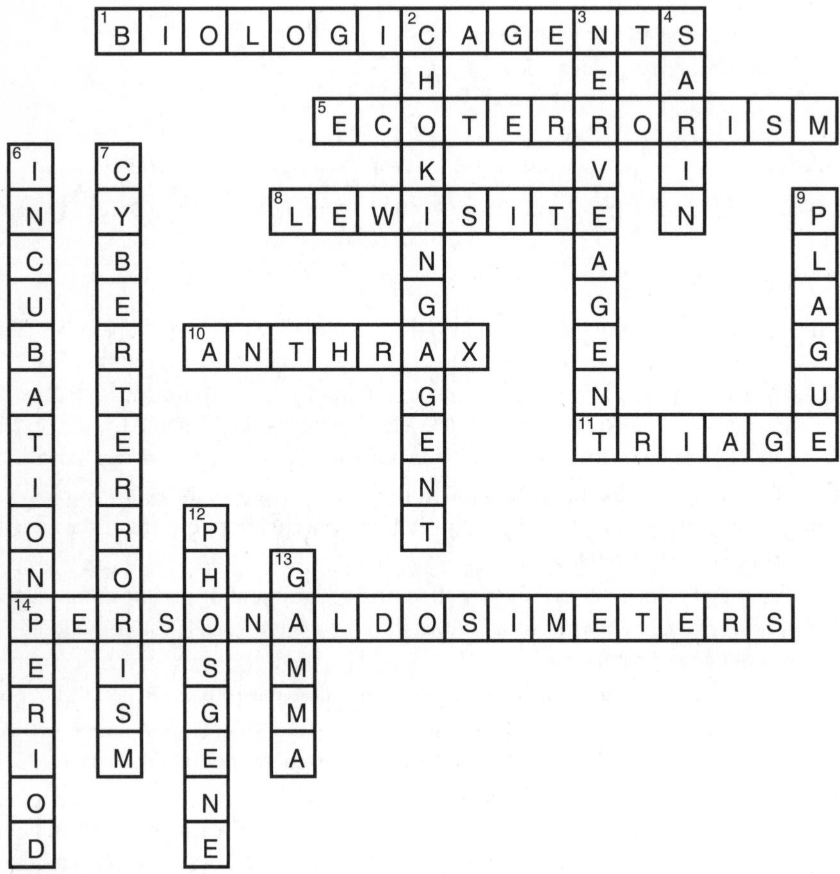

Fire Alarms

1. The fire service role includes emergency medical services (EMS), hazardous materials mitigation, and technical rescue as well as fire suppression. All of these functions will probably be needed when a terrorist incident occurs. Terrorism presents challenges to the fire service on a scale that has never previously been experienced in North America; it also presents an unparalleled threat to the lives of fire fighters and emergency responders. The terrorist threat requires fire fighters to work closely with local, state, and federal law enforcement agencies; emergency management agencies; allied health agencies; and the military. It is critical that all of these agencies work together in a coordinated and cooperative manner. All emergency responders and law enforcement agencies must be prepared to face a wide range of potential situations.
2. Unless the cause of an explosion is known to be accidental, fire fighters at the scene should always consider the possibility that an explosive device was detonated. The first priority should be to ensure the safety of the scene. Fire fighters should also consider the possibility that a secondary device may be in the vicinity. Responders should quickly survey the area for any suspicious bags, packages, or other items. It is also possible that chemical, biologic, or radiological agents may be involved in a terrorist bombing. Qualified personnel with monitoring instruments should be assigned to check the area for potential contaminants. These precautions should be implemented immediately. The initial size-up should also include an assessment of hazards and dangerous situations.

   The stability of any building involved in the explosion must be evaluated before anyone is permitted to enter. Entering an unstable area without the proper training and equipment may complicate rescue and recovery efforts.

Chapter 7: Mission-Specific Competencies: Personal Protective Equipment

Matching

1. B (page 161)
2. E (page 163)
3. D (page 175)
4. A (page 173)
5. G (page 163)
6. C (page 162)
7. F (page 162)
8. I (page 164)
9. J (page 165)
10. H (page 171)

Multiple Choice

1. C (page 162)
2. C (page 163)
3. B (page 163)
4. D (page 162)
5. B (page 171)
6. C (page 162)
7. D (page 163)
8. D (page 162)
9. A (page 163)
10. B (page 163)

Labeling

A. Vapor-protective clothing (page 163)

B. High temperature–protective clothing (page 162)

C. Liquid splash–protective clothing (pages 163–164)

Vocabulary

1. **Level A ensemble:** Personal protective equipment that is used when the type and atmospheric concentration of substances requires a high level of respiratory protection but less skin protection. The kind of gloves and boots worn depends on the identified chemical. (page 165)
2. **Allied professional:** A person with unique skills, knowledge, and/or abilities who may be called upon to assist hazardous materials responders. Examples of allied professionals may include a Certified Industrial Hygienist (CIH), Certified Safety Professional (CSP), Certified Health Physicist (CHP), or similar credentialed or competent individuals as determined by the authority having jurisdiction. (page 175)
3. **Dehydration:** An excessive loss of body water. Signs and symptoms of dehydration may include increasing thirst, dry mouth, weakness or dizziness, a darkening of the urine, or a decrease in the frequency of urination. (page 175)
4. **Donning:** The process of putting on an ensemble of PPE. (page 165)
5. **High temperature–protective equipment:** A type of personal protective equipment that shields the wearer during short-term exposures to high temperatures. Sometimes referred to as a proximity suit, this type of equipment allows the properly trained fire fighter to work in extreme fire conditions. It is not designed to protect against hazardous materials or weapons of mass destruction. (page 162)

Fill-in

1. least (page 173)
2. Liquids (page 163)
3. Chemical (page 162)
4. single (page 165)
5. heat-exchange (page 178)
6. Gamma (page 162)
7. TRACEMP (page 160)
8. Alpha (page 161)
9. 8 to 16 ounces (page 178)
10. HAZWOPER (page 164)

True/False

1. T (page 161)
2. T (page 161)
3. F (page 175)
4. F (page 163)
5. F (page 163)
6. T (page 159)
7. T (page 159)
8. T (page 161)
9. F (page 162)
10. T (page 163)

Short Answer

1. *TRACEMP:* Thermal; Radiological; Asphyxiating; Chemical; Etiological/Biological; Mechanical; Psychogenic. (page 160)
2. *Level A protection:* Personal protective equipment provides protection against vapors, gases, mists, and even dusts. The highest level of protection, it requires a totally encapsulating suit that includes self-contained breathing apparatus.

   Level B protection: Personal protective equipment is used when the type and atmospheric concentration of substances have been identified. It generally requires a high level of respiratory protection but less skin protection (chemical-protective coveralls and clothing, chemical protection for shoes, gloves, and self-contained breathing apparatus outside of a nonencapsulating chemical-protective suit).

   Level C protection: Personal protective equipment is used when the type of airborne substance is known, the concentration is measured, the criteria for using air-purifying respirators are met, and skin and eye exposure is unlikely. It consists of standard work clothing with the addition of chemical-protective clothing, chemically resistant gloves, and a form of respirator protection.

   Level D protection: Personal protective equipment is used when the atmosphere contains no known hazard, and work functions preclude splashes, immersion, or the potential for unexpected inhalation of or contact with hazardous levels of chemicals. It is primarily a work uniform that includes coveralls and affords minimal protection. (pages 165–173)
3. Allied professionals might include a Certified Industrial Hygienist (CIH), Certified Safety Professional (CSP), Certified Health Physicist (CH), or similar credentialed or competent individuals as determined by the authority having jurisdiction. (page 175)

Word Fun

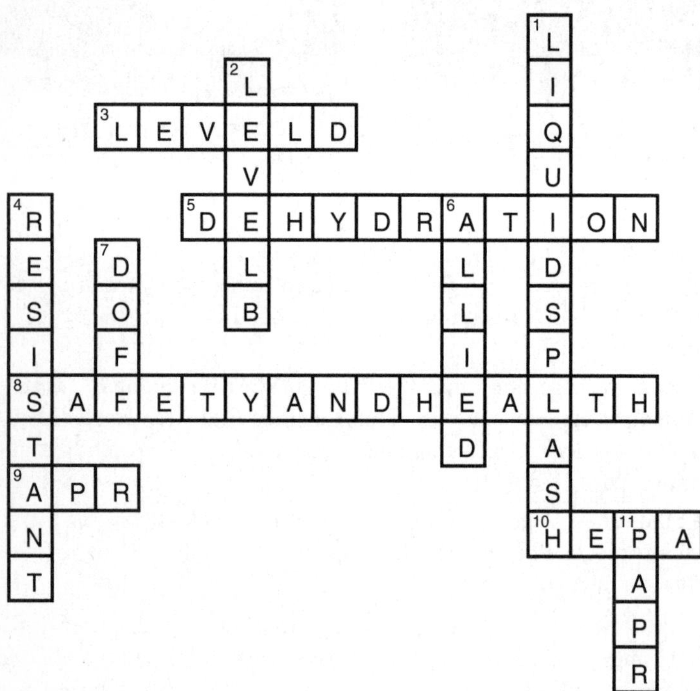

Fire Alarms

1. The recommended PPE for Level B protection includes the following components:
   - SCBA or SAR
   - Chemical-resistant clothing
   - Inner and outer chemical-resistant gloves
   - Chemical-resistant safety boots/ shoes

- Hard hat
- Two-way radio

Optional PPE for Level B protection includes the following components:

- Coveralls
- Long cotton underwear
- Disposable gloves and boot covers

2. (a) Heat exhaustion is a mild form of shock that occurs when the circulatory system begins to fail because the body is unable to dissipate excessive heat and becomes overheated.

(b) Although heat exhaustion is not an immediately life-threatening condition, the affected individual should be removed at once from the source of heat, rehydrated with electrolyte solutions, and kept cool. If not properly treated, heat exhaustion may progress to heat stroke.

Skill Drills

Skill Drill 7-1: Donning a Level A Ensemble

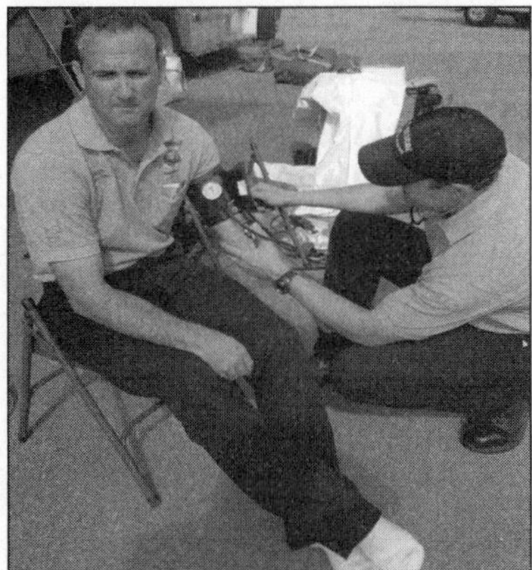

1. Conduct a pre-entry briefing, medical monitoring, and equipment inspection.

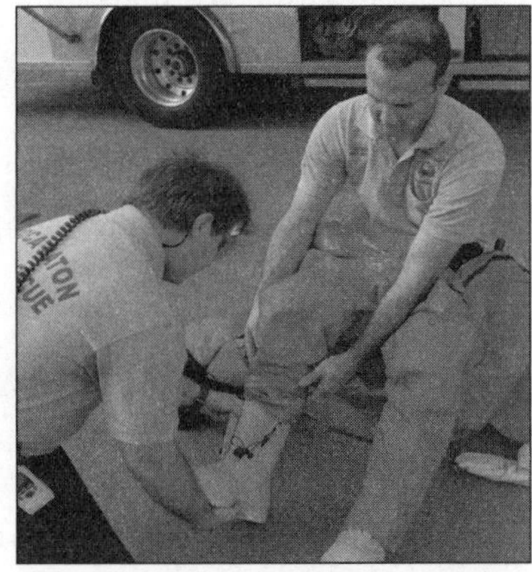

2. While seated, pull on the suit to waist level; pull on the chemical boots over the top of the chemical suit. Pull the suit boot covers over the tops of the boots.

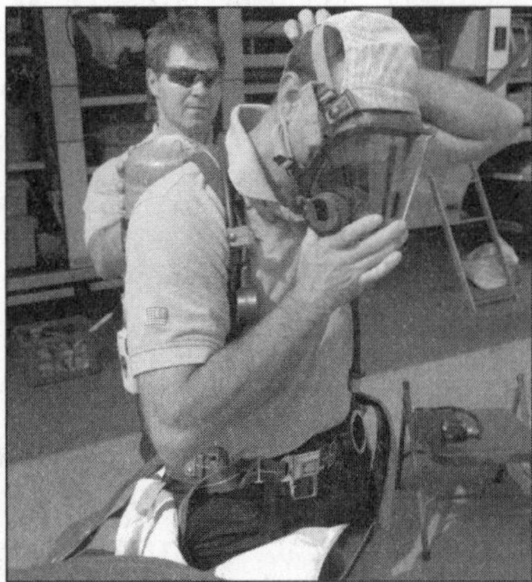

3. Stand up and don the SCBA frame and SCBA face piece, but do not connect the regulator to the face piece.

4. Place the helmet on the head.

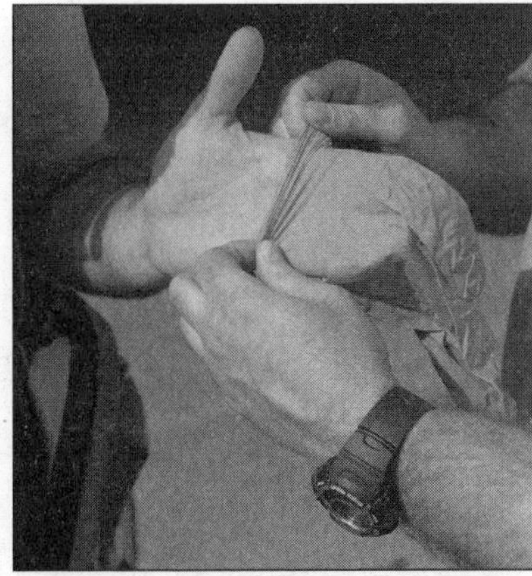

5. Don the inner gloves.

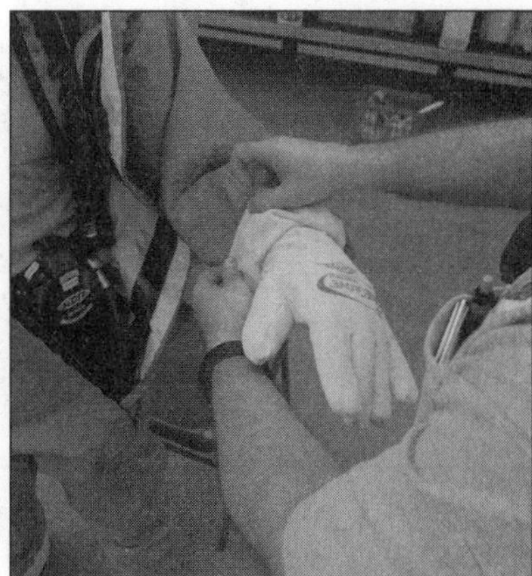

6. Don the outer chemical gloves (if required by the manufacturer's specifications). With assistance, complete donning the suit by placing both arms in the suit, pulling the expanded back piece over the SCBA, and placing the chemical suit over the head.

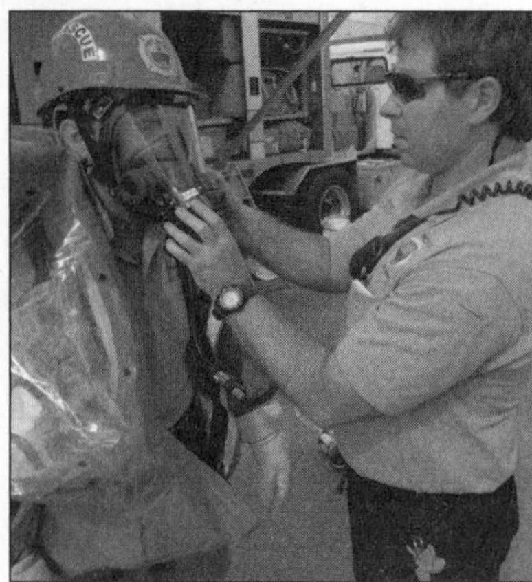

7. Instruct the assistant to connect the regulator to the SCBA face piece and ensure air flow.

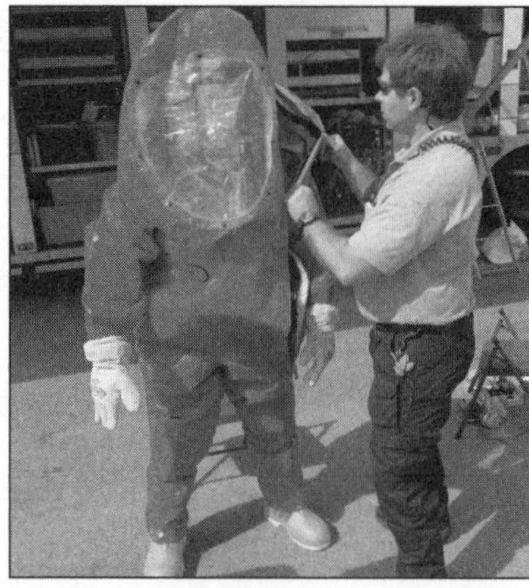

8. Instruct the assistant to close the chemical suit by closing the zipper and sealing the splash flap.

9. Review hand signals and indicate that you are okay.

Skill Drill 7-2: Doffing a Level A Ensemble

1. After completing decontamination, proceed to the clean area for suit doffing. Pull the hands out of the **outer gloves** and arms from the sleeves, and cross the arms in front inside the suit.
2. Instruct the assistant to open the **chemical splash** flap and suit zipper.
3. Instruct the assistant to begin at the head and roll the suit **down** and **away** until the suit is below waist level.
4. Instruct the assistant to complete rolling the suit from the waist to the ankles; step out of the attached **chemical boots** and suit.
5. Doff the SCBA frame. The **face piece** should be kept in place while the SCBA frame is doffed.
6. Take a deep breath and doff the SCBA face piece; carefully peel off the inner gloves, and walk away from the clean area. Go to the rehabilitation area for **medical monitoring**, rehydration, and personal decontamination shower.

Skill Drill 7-3: Donning a Level B Nonencapsulated Chemical-Protective Clothing Ensemble

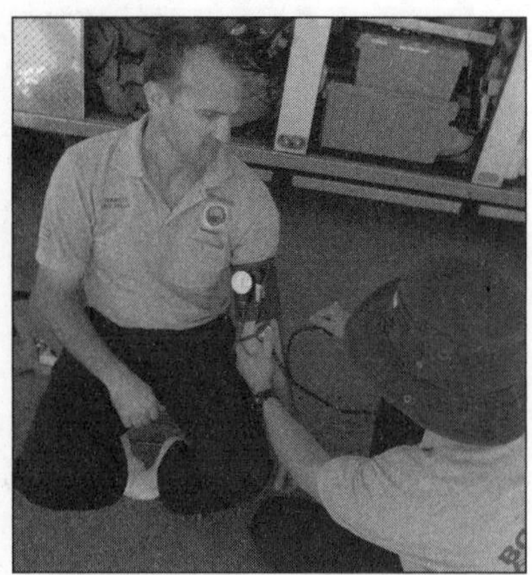

1. Conduct a pre-entry briefing, medical monitoring, and equipment inspection.

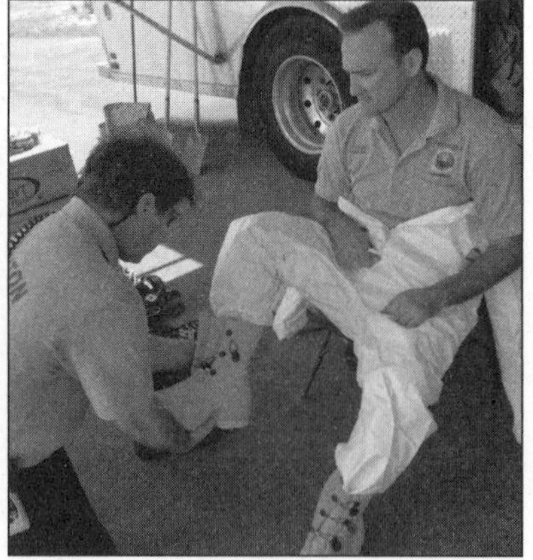

2. While seated, pull on the suit to waist level; pull on the chemical boots over the top of the chemical suit. Pull the suit boot covers over the tops of the boots.

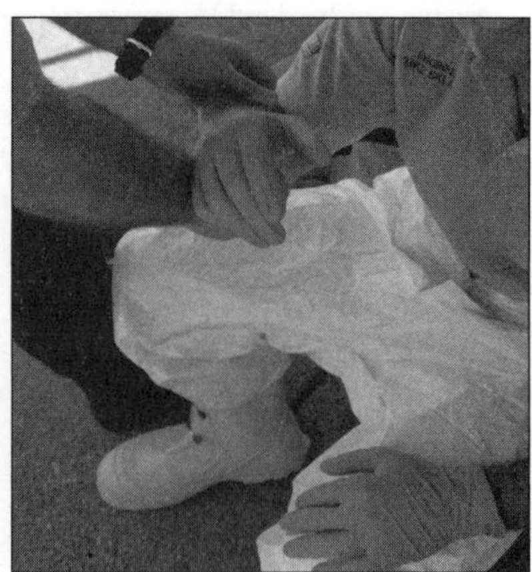

3. Don the inner gloves.

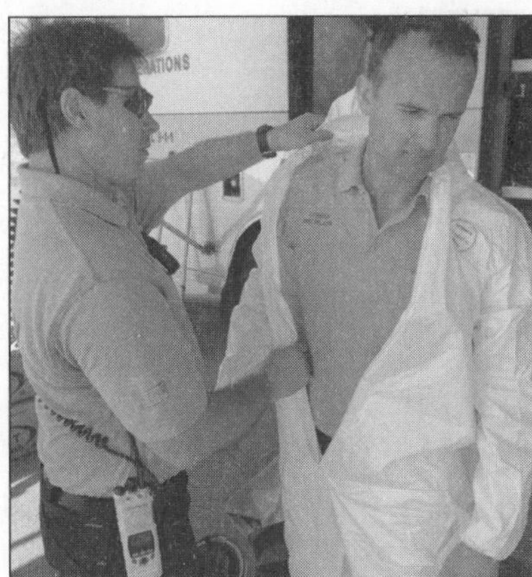

4. With assistance, complete donning the suit by placing both arms in the suit and pulling the suit over the shoulders. Instruct the assistant to close the chemical suit by closing the zipper and sealing the splash flap.

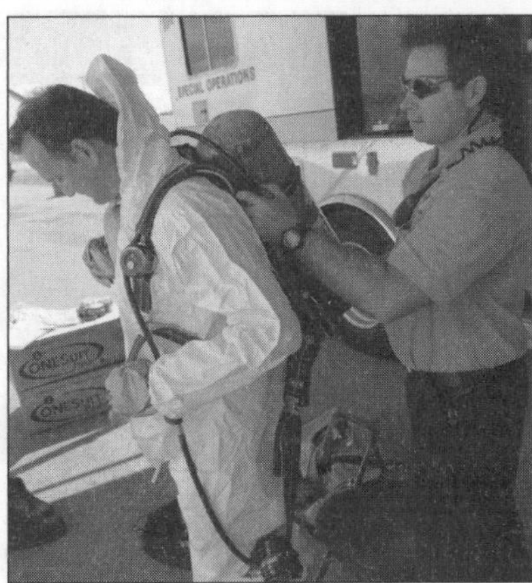

5. Don the SCBA frame and SCBA face piece, but do not connect the regulator to the face piece.

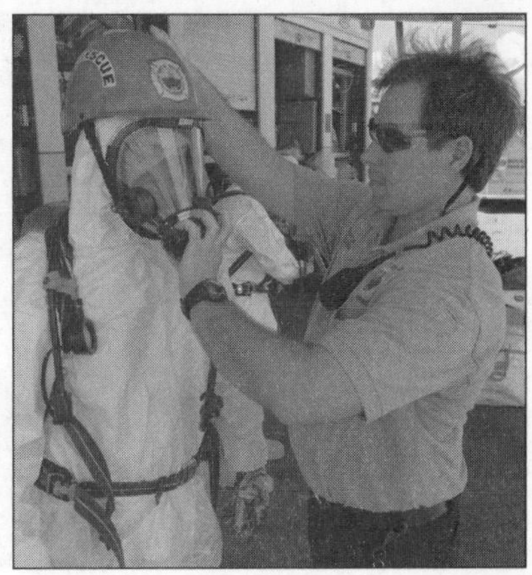

6. With assistance, pull the hood over the head and SCBA face piece. Place the helmet on the head. Put on the outer gloves. Instruct the assistant to connect the regulator to the SCBA face piece and ensure you have air flow.

Skill Drill 7-4: Doffing a Level B Nonencapsulated Chemical-Protective Clothing Ensemble

1. After completing decontamination, proceed to the **clean area** for suit doffing. Stand and doff the SCBA frame. Keep the face piece in place.
2. Instruct the assistant to open the chemical splash flap and open the **suit zipper**.
3. Remove your hands from the outer gloves and arms from the sleeves of the suit. Cross your arms in front **inside** the suit. Instruct the assistant to begin at the head and roll the suit down and away until the suit is below waist level.
4. Sit down and instruct the assistant to complete rolling down the suit to the **ankles**; step out of the attached chemical boots and suit.
5. Stand and doff the SCBA face piece and **helmet**.
6. Carefully peel off the inner gloves, and walk away from the clean area. Go to the rehabilitation area for medical monitoring, rehydration, and **personal decontamination shower**.

Skill Drill 7-5: Donning a Level C Chemical-Protective Clothing Ensemble

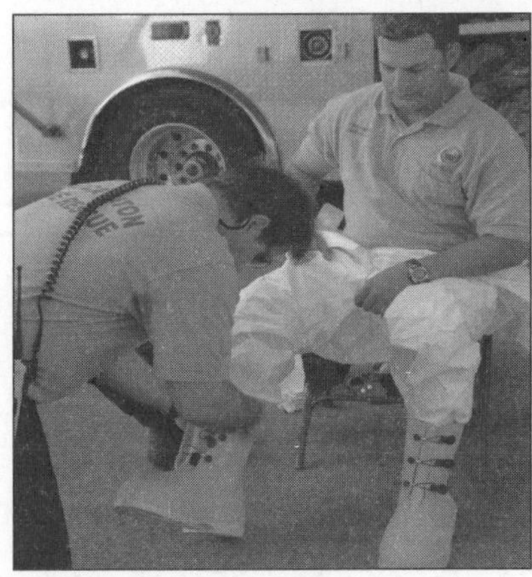

1. Conduct a pre-entry briefing, medical monitoring, and equipment inspection. While seated, pull on the suit to waist level; pull on the chemical boots over the top of the chemical suit. Pull the suit boot covers over the tops of the boots.

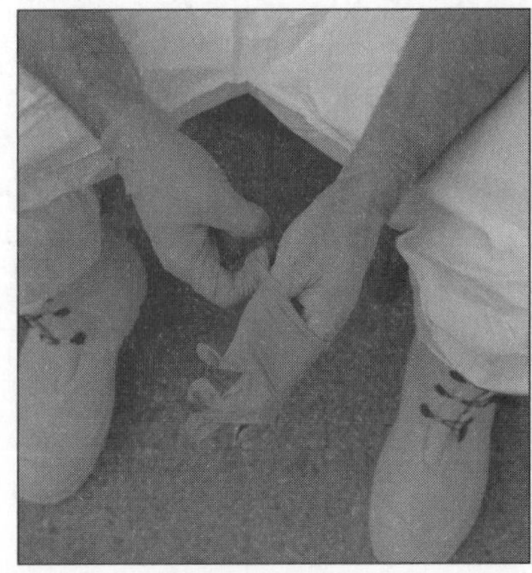

2. Don the inner gloves.

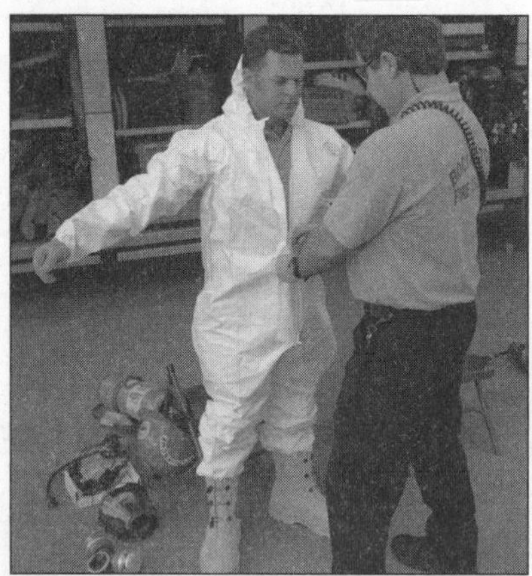

3. With assistance, complete donning the suit by placing both arms in the suit and pulling the suit over the shoulders. Instruct the assistant to close the chemical suit by closing the zipper and sealing the splash flap.

4. Don APR/PAPR. Pull the hood over the head and APR/PAPR face piece. Place the helmet on the head. Pull on the outer gloves. Review hand signals and indicate that you are okay.

Skill Drill 7-6: Doffing a Level C Chemical-Protective Clothing Ensemble

1. After completing **decontamination**, proceed to the clean area. As with Level B, the assistant opens the chemical splash flap and suit zipper. Remove the hands from the outer gloves and arms from the sleeves. Instruct the assistant to begin at the head and roll the suit down below waist level. Instruct the assistant to complete rolling down the suit and take the **chemical boots** and suit away. The assistant helps remove the inner gloves. Remove APR/PAPR. Remove the helmet.
2. Go to the **rehabilitation** area for medical monitoring, rehydration, and personal decontamination shower.

Skill Drill 7-7: Donning a Level D Chemical-Protective Clothing Ensemble

1. Conduct a pre-entry briefing and equipment inspection. Don the Level D suit. Don the boots. Don safety glasses or **chemical goggles**. Don a hard hat. Don gloves, a face **shield**, and any other required equipment.

Chapter 8: Mission-Specific Competencies: Technical Decontamination

Matching

1. F (page 188)
2. D (page 191)
3. G (page 192)
4. B (page 190)
5. C (page 192)
6. A (page 191)
7. E (page 190)
8. I (page 191)
9. J (page 191)
10. H (page 190)

Multiple Choice

1. A (page 188)
2. B (page 190)
3. D (page 188)
4. A (page 190)
5. C (page 192)
6. B (page 194)
7. A (page 194)
8. B (page 194)
9. C (page 194)
10. D (page 192)

Labeling

1. Physical methods of technical decontamination.

A. Adsorption

B. Absorption

Vocabulary

1. **Decontamination team:** The team responsible for reducing and preventing the spread of contaminants from persons and equipment used at a hazardous materials incident. Members of this team establish the decontamination corridor and conduct all phases of decontamination. (page 189)
2. **Contamination:** The process of transferring a hazardous material from its source to people, animals, the environment, or equipment, any of which may act as a carrier. (page 188)
3. **Adsorption:** The process of adding a material such as sand or activated carbon to a contaminant, which then adheres to the surface of the material and allows for collection of the contaminated material. (page 190)
4. **Solidification:** The process of chemically treating a hazardous liquid so as to turn it into a solid material, thereby making the material easier to handle. (page 192)
5. **Sterilization:** A process utilizing heat, chemical means, or radiation to kill microorganisms. (page 192)

Fill-in

1. decontamination corridor (page 188)
2. Gross decontamination (page 188)
3. evaporate (page 191)
4. disinfection (page 191)
5. dilution (page 191)
6. Isolation, disposal (page 192)
7. adsorption (page 191)
8. Technical decontamination (page 188)
9. cold (page 192)
10. identify (page 189)

True/False

1. F (page 188)
2. F (pages 188, 189)
3. T (page 190)
4. T (page 194)
5. T (page 189)

Short Answer

1. *Emergency decontamination:* Used in potentially life-threatening situations to rapidly remove the bulk of the contamination from an individual, regardless of the presence or absence of a technical decontamination corridor.

   Gross decontamination: Like emergency decontamination, gross decontamination aims to significantly reduce the amount of surface contaminant by delivery of a continuous shower of water and removal of outer clothing or PPE. It differs from emergency decontamination, however, in that gross decontamination is controlled through the decontamination corridor.

   Technical decontamination: Performed after gross decontamination and is a more thorough cleaning process. Technical decontamination may involve several stations or steps. During this type of decontamination, multiple personnel (the decontamination team) typically use brushes to scrub and wash the person or object to remove contaminants.

   Mass decontamination: Often used in incidents involving unknown agents or in the case of a contamination of large groups of people. It takes place in the field and is a way of quickly performing gross decontamination on a large number of victims who have escaped from a hazardous materials incident. (pages 188, 189)

Word Fun

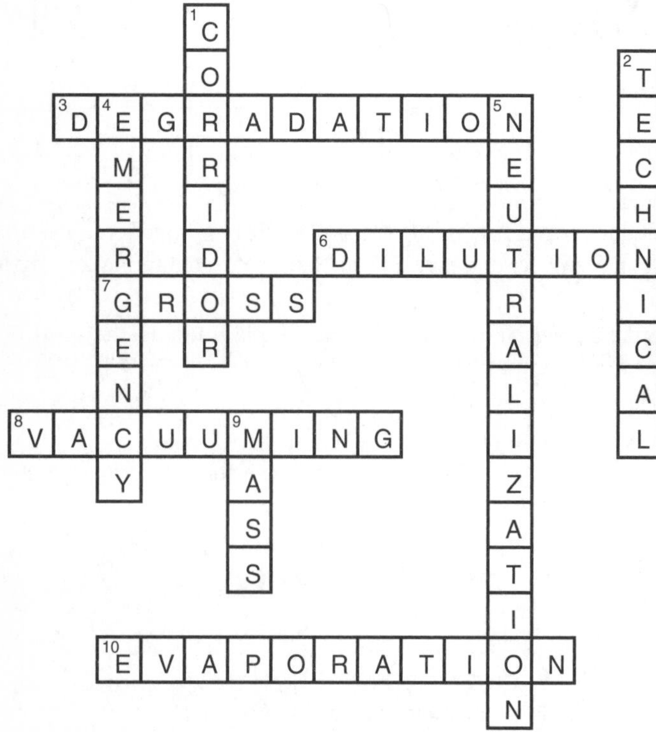

Fire Alarms

1. Ensure that you have the appropriate PPE to protect against the chemical threat. Stay clear of the product, and do not make physical contact with it. Make an effort to contain runoff by directing victims out of the hazard zone and into a suitable location for decontamination. Flush the victim to remove the product from the victim's clothing. Instruct and assist the victim in removing contaminated clothing. Flush the contaminated victim. Assist or obtain medical treatment for the victim, and arrange for the victim's transport.
2. Alternative decontamination procedures include the following techniques:
   - Absorption
   - Adsorption
   - Dilution
   - Disinfection
   - Disposal
   - Solidification
   - Emulsification
   - Vapor dispersion
   - Removal
   - Vacuuming

Skill Drills

Skill Drill 8-1: Performing Technical Decontamination on a Responder

1. Drop any tools and equipment.

2. Perform gross decontamination, if necessary.

3. Perform technical decontamination. Wash and rinse responder one to three times.

4. Remove outer hazardous materials–protective clothing.

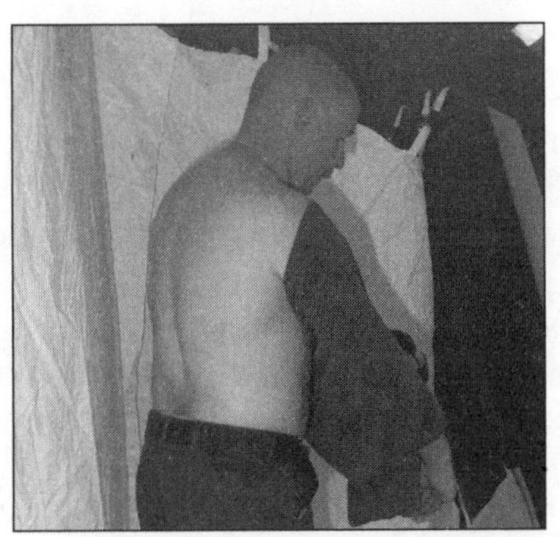

5. Remove personal clothing. Proceed to the rehabilitation area for medical monitoring, rehydration, and personal decontamination shower.

Chapter 9: Mission-Specific Competencies: Mass Decontamination

Matching

1. C (page 206) **2.** D (page 210) **3.** A (page 210) **4.** B (page 210) **5.** E (page 210)

Multiple Choice

1. D (page 204) **2.** C (page 205) **3.** C (page 206) **4.** D (page 210) **5.** B (page 210)

Labeling

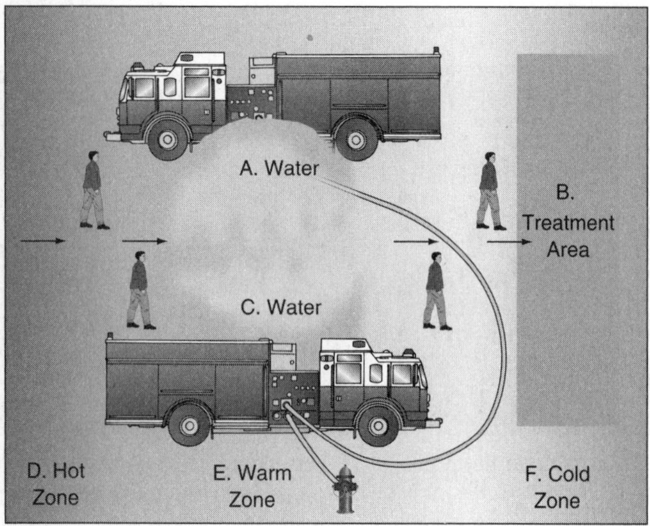

Vocabulary

1. **Mass decontamination:** The physical process of reducing or removing surface contaminants from large numbers of victims in potentially life-threatening situations in the fastest time possible. (page 204)
2. **Dilution:** The process of adding a substance—usually water—in an attempt to weaken the concentration of another substance. (page 210)
3. **Isolation and disposal:** A two-step removal process for items that cannot be properly decontaminated. The contaminated article is removed and isolated in a designated area and then packaged in a suitable container. The second step involves transport of the item to an approved facility, where it is either incinerated or buried in a hazardous waste landfill. (page 210)

Fill-in

1. 70°F (21°C) (page 207)
2. decontamination (page 208)
3. airway (page 208)
4. 15 (page 210)
5. valuables, clothing (page 215)

True/False

1. F (page 204)
2. T (page 205)
3. F (page 205)
4. F (page 206)
5. F (page 207)
6. T (page 208)
7. F (page 208)
8. T (page 210)
9. F (page 214)
10. T (page 213)

Short Answer

1. The NFPA 704 marking system is designed for fixed-facility use. Locations where the NFPA 704 system might be used include on the outsides of buildings, on doorways to chemical storage areas, and on fixed storage tanks. (page 213)
2. Placards are diamond-shaped indicators for use in transport. Locations where a placard system might be used include on all four sides of highway transport vehicles, railroad tank cars, and other forms of transportation carrying hazardous materials. (page 212)

Word Fun

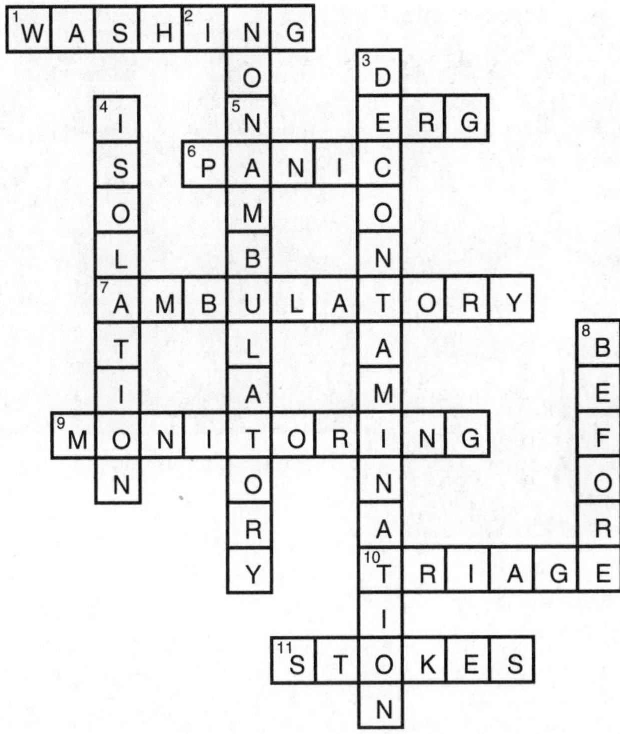

Fire Alarms

1. Mass decontamination can be performed in a number of ways. In the fire service, this operation is sometimes accomplished by placing two fire apparatus side by side. Fog-type nozzles are attached to the pumpers positioned opposite of each other, and a fog pattern is used to douse the victims as they walk between the two pieces of apparatus. Pump pressures are set so that the fog patterns are effective but not overwhelming (usually between 30 and 50 psi). An aerial ladder device (either in conjunction with the pumpers or as a stand-alone unit) can also provide a complete overhead spray pattern using pre-plumbed waterways or other configurations of hose lines. (page 205)
2. The standard thinking regarding decontamination—using minimal amounts of water, and recovering the potentially contaminated runoff—becomes a secondary objective when mass decontamination is implemented. If lives are at stake, controlling runoff is not the responders' main concern. (page 205)

Skill Drills

Skill Drill 9-1: Performing Mass Decontamination on Ambulatory Victims

1. Ensure you have the appropriate PPE to protect against the chemical threat. Stay clear of the product and do not make physical contact with it. Make an effort to contain **runoff** by directing victims out of the **hazard** **zone** and into a suitable location.
2. Set up the appropriate type of mass decontamination system based on the type of **apparatus**, equipment, and/or system available.
3. Instruct victims to **remove** their contaminated clothing and walk through the decontamination corridor. Flush the contaminated victims with **water**.
4. Direct the contaminated victims to the **triage** area.

Skill Drill 9-2: Performing Mass Decontamination on Nonambulatory Victims

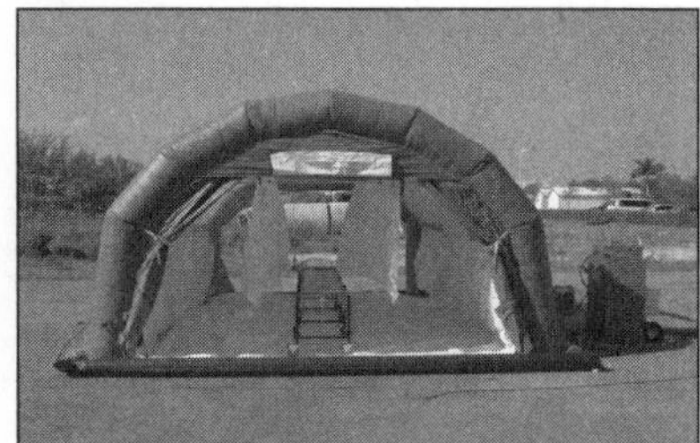

1. Set up the appropriate type of mass decontamination system based on the type of equipment available.

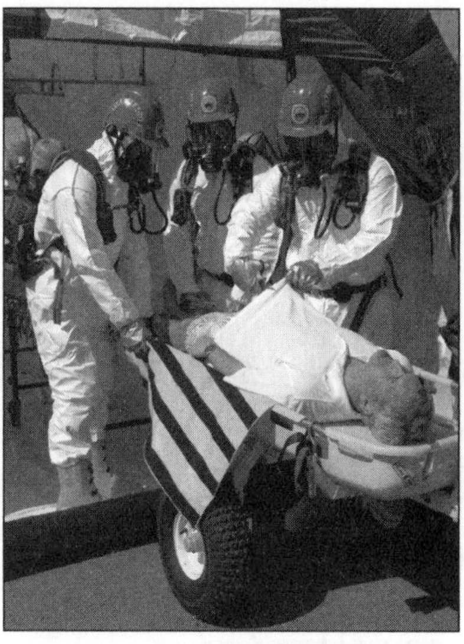

2. Ensure you have the appropriate PPE to protect against the chemical threat. Remove the victim's clothing. Do not leave any clothing underneath the victim; these items may wick the contamination to the victim's back and hold it there, potentially worsening the exposure.

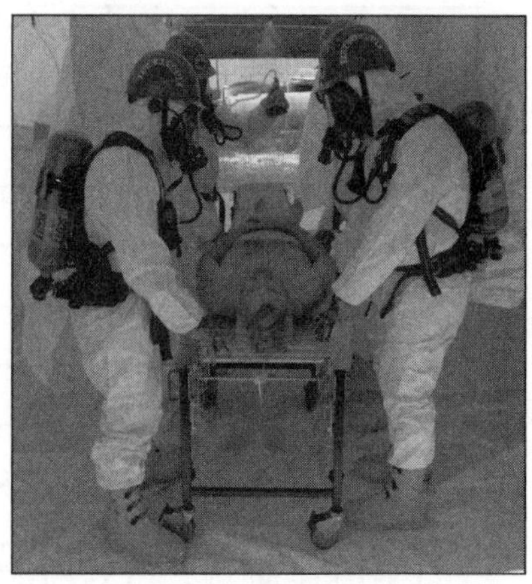

3. Flush the contaminated victims with water.

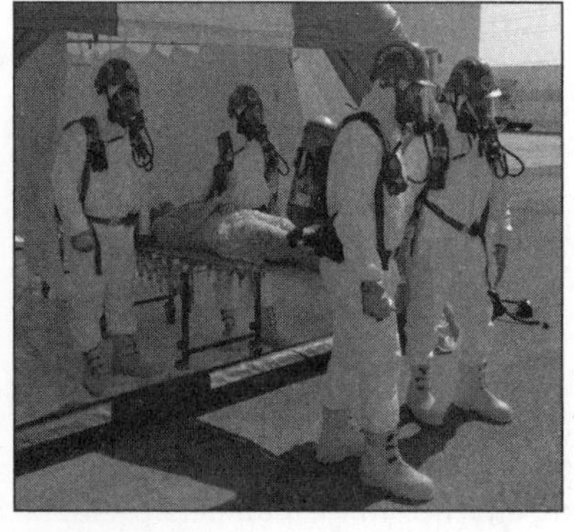

4. Move the victim to a designated triage area for medical evaluation.

Chapter 10: Mission-Specific Competencies: Evidence Preservation and Sampling

Matching

1. B (page 223) **2.** C (page 225) **3.** E (page 229) **4.** A (page 225) **5.** D (page 225)

Multiple Choice

1. A (page 225) **2.** C (page 225) **3.** D (page 225) **4.** A (page 226) **5.** B (page 224)

Vocabulary

1. **Chain of custody:** A record documenting the identities of any personnel who handled the evidence, the date and time that contact occurred or the evidence was transferred from one person to another, and the reason for the handling or transfer of evidence. (page 229)
2. **Investigative authority:** The agency that has the legal jurisdiction to enforce a local, state, or federal law or regulation. It is the most appropriate law enforcement organization to ensure the successful investigation and prosecution of a case. (page 225)
3. **Demonstrative evidence:** Materials used to demonstrate a theory or explain an event. (page 225)

Fill-in

1. Physical (page 225)
2. direct (page 226)
3. evidence preservation (page 227)
4. contaminated (page 227)
5. explosive ordnance disposal (EOD) (page 232)
6. "Tag in/tag out" (page 232)
7. certified, sterile (page 234)
8. demonstrative (page 225)
9. unified command (page 225)
10. Responders (page 229)

True/False

1. T (page 224)
2. F (page 225)
3. T (page 224)
4. F (page 227)
5. T (page 227)

Short Answer

1. The 12-step process recommended by the FBI regarding the collection or sampling of evidence includes:
   (a) Preparation
   (b) Approach the scene
   (c) Secure and protect the scene
   (d) Initiate a preliminary survey
   (e) Evaluate physical evidence possibilities
   (f) Prepare a narrative description
   (g) Depict the scene photographically
   (h) Prepare a diagram or sketch of the scene
   (i) Conduct a detailed search
   (j) Record and collect physical evidence
   (k) Conduct the final survey
   (l) Release the scene (page 230)

2. Indicators that legitimate toxic industrial chemicals were released intentionally include:
   (a) opened valves
   (b) punctures or cuts in containment vessels
   (c) cut chains, locks, or other safety devices that would normally have been in place to contain the hazard (page 224)

Word Fun

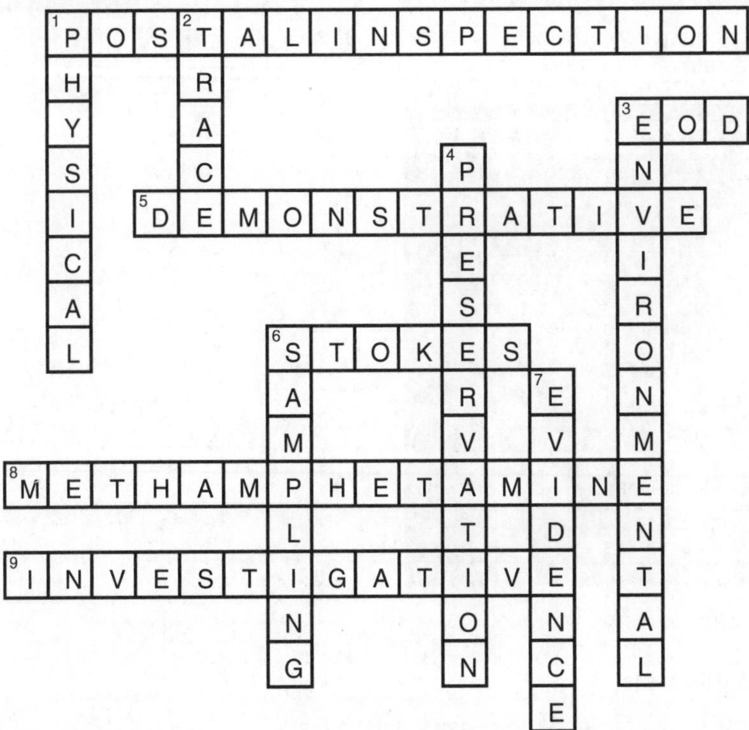

Fire Alarms

1. Indicators of an illicit laboratory may include efforts taken to conceal the presence of the lab, such as fences, excessive window coverings, or enhanced ventilation and air filtration systems. Alarms and counter-surveillance systems may also be found at such locations. Other, more obvious indicators could include the presence of chemical storage cylinders and glass bottles, laboratory glassware, electric heat sources, and Bunsen burners. (page 224)
2. Environmental crimes include the intentional release or disposal of hazardous materials, their byproducts, and waste into the environment. These releases may occur into the air, into the ground, or into natural or human-made water systems. Indicators that such crimes have occurred may include containers (labeled or unlabeled) that have been discarded at the site, staining or odors near street drainage systems, and dead or dying plants, insects, or animals in the nearby area. (page 224)

Skill Drills

Skill Drill 10-1: Collecting and Processing Evidence

1. Take photographs of the evidence.

2. Sketch, mark, and label the location of the evidence.

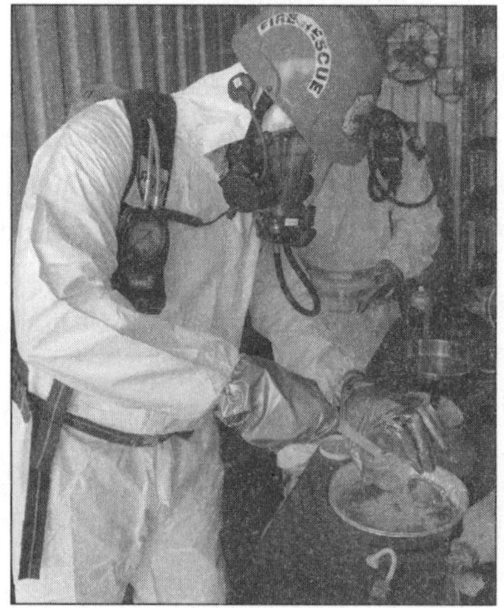

3. Place the evidence in the appropriate container.

4. Tag the evidence with labels.

5. Document your findings.

Skill Drill 10-2: Securing, Characterizing, and Preserving the Scene

1. Observe the scene for certain characteristics that could lead to the discovery of **evidence**.
2. Place caution tape around the scene to limit **access**.
3. **Preserve** suspected evidence by protecting it from being disturbed.

Skill Drill 10-6: Collecting Samples Utilizing Equipment and Preventing Secondary Contamination

1. The sampler obtains two samples: one sample to use for **field screening** and another sample to preserve as **evidence**.
2. The **assistant** holds open the evidence package or container so the sampler can place the evidence inside without cross-contaminating the evidence.

Skill Drill 10-7: Documenting Evidence

1. The **documenter** photographs and/or videotapes the sampling and collection process.
2. The documenter makes notes about the name of the person **collecting** or **sampling** the evidence, the physical location of the agent, the state of the agent, the **quantity** present, the time of the sample collection, and the size and **condition** of the container.

Skill Drill 10-8: Evidence Labeling, Packaging, and Decontamination

1. Seal the initial container with tape, and place your **initials** on the tape or seal. This step will prevent **tampering**.
2. Place the initial container in a **secondary** container and label it with a unique **exhibit** number, the name of the person who collected the item, and the location, time, and date of the evidence collection.

Chapter 11: Mission-Specific Competencies: Product Control

Matching

1. C (page 253)
2. A (page 244)
3. B (page 244)
4. F (page 250)
5. E (page 250)
6. D (page 253)
7. J (page 251)
8. I (page 248)
9. H (page 246)
10. G (page 245)

Multiple Choice

1. A (page 246)
2. C (page 250)
3. B (page 256)
4. A (page 255)
5. D (page 246)
6. A (page 244)
7. C (page 253)
8. C (page 245)
9. D (page 246)
10. A (page 251)
11. B (page 250)
12. D (page 253)
13. C (page 256)
14. B (page 256)
15. B (page 259)

Vocabulary

1. **Exposures:** People, property, structures, or parts of the environment that are subject to influence, damage, or injury as a result of contact with a hazardous material. The amount of exposures that remain is determined both by the location of the incident and by the amount of progress that has been made in protecting those exposures via isolation and other indirect responses. Incidents in urban areas will likely have more exposures and, therefore, will likely need more resources to protect those exposures from the hazardous materials. (page 245)
2. **Underflow dam:** A method of containing materials that are lighter than water. (page 247)
3. **Recovery phase:** The stage of a hazardous materials incident after the imminent danger has passed, when clean-up and the return to normalcy have begun. (page 256)

Fill-in

1. vapor dispersion (page 253)
2. disperse (page 253)
3. evaporate (page 245)
4. unnecessary (page 245)
5. remote shutoffs (page 251)
6. Retention (page 251)
7. complete dam (page 246)
8. dissolve (page 255)
9. liquid fires (page 253)
10. Containment (page 244)

True/False

1. T (page 244)
2. T (page 246)
3. F (page 245)
4. F (page 255)
5. T (page 245)
6. T (page 256)
7. F (page 252)
8. T (page 251)
9. T (page 250)
10. F (page 251)

Short Answer

1. Aqueous film-forming foam (AFFF); fluoroprotein foam; protein foam; high-expansion foam; alcohol-resistant concentrates (page 255)

Word Fun

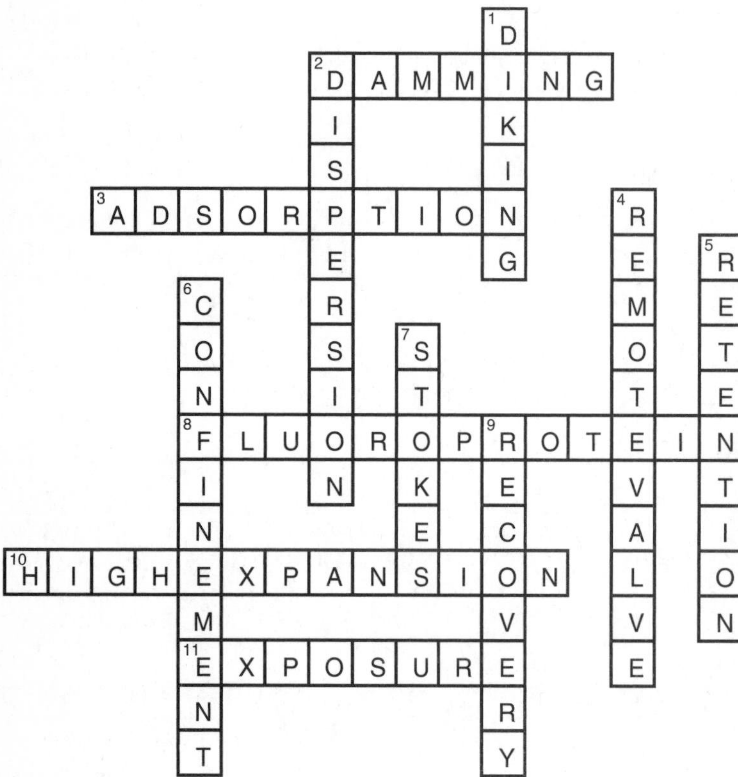

Fire Alarms

1. Collect the basic materials: absorbents, adsorbents, absorbent pads, and absorbent booms. Decide which material is best suited for use with the spilled product, and assess the location of the spill. Stay clear of any spilled product. Apply the appropriate material to control and contain the spilled material. Maintain materials, and take appropriate steps for their disposal.
2. (a) AFFF, protein foam, and fluoroprotein foam
   (b) Foam should be gently applied or bounced off another adjacent object so that it flows down across the liquid and does not directly upset the burning surface. Foam can also be applied in a rain-down method by directing the stream into the air over the material and letting the foam fall gently, much as rain would.

Skill Drills

Skill Drill 11-1: Using Absorption/Adsorption to Manage a Hazardous Materials Incident

1. Decide which **material** is best suited for use with the spilled product. Access the location of the spill, and stay clear of any spilled product.
2. Use detection and monitoring **devices** as well as reference sources to identify the product. Apply the appropriate material to control the spilled product.
3. Maintain **control** of the absorbent/adsorbent materials, and take appropriate steps for their disposal.

Skill Drill 11-4: Constructing a Dike

1. Determine the best location for the dike. If necessary, dig a depression in the ground 6" to 8" (15 cm to 20 cm) deep. Ensure that plastic will not **react** adversely with the spilled chemical. Use plastic to line the bottom of the depression, and allow for sufficient plastic to cover the dike wall.

2. Build a short wall with **sandbags** or other available materials.
3. Complete the dike installation, and ensure that its **size** will contain the spilled product.

Skill Drill 11-8: Using Vapor Dispersion to Manage a Hazardous Materials Incident

1. Determine the viability of a dispersion operation. Use the appropriate monitoring instrument to determine the boundaries of a safe work area. Ensure that **ignition sources** in the area have been removed or controlled.
2. Apply water from a distance to disperse vapors. **Monitor** the environment until the vapors have been adequately dispersed.

Skill Drill 11-12: Performing the Bounce-Off Method of Applying Foam

1. Open the nozzle and **test** to ensure that foam is being produced.
2. Move within a safe range of the product, and open the nozzle. Direct the stream of foam onto a solid structure, such as a wall or metal tank, so that the foam is directed off the object and onto the pool of product. Allow the foam to flow across the top of the pool of product until it is completely covered. Be aware that the foam may need to be banked off of several areas of the solid object so as to **extinguish** the burning product.

Chapter 12: Mission-Specific Competencies: Victim Rescue and Recovery

Matching

1. C (page 267)
2. D (page 274)
3. A (page 266)
4. B (page 266)
5. E (page 268)

Multiple Choice

1. C (page 269)
2. D (page 266)
3. D (page 276)
4. A (page 277)
5. B (page 266)

Labeling

1. Search and rescue/recovery equipment.

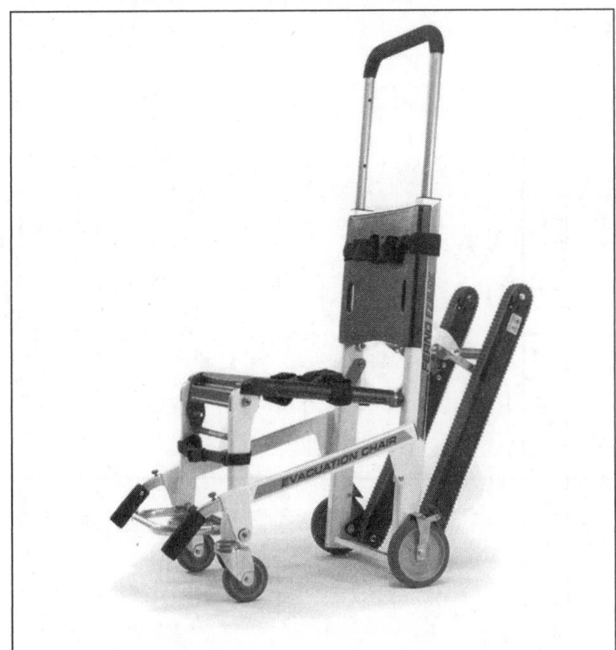

A. Evacuation chair (page 272)

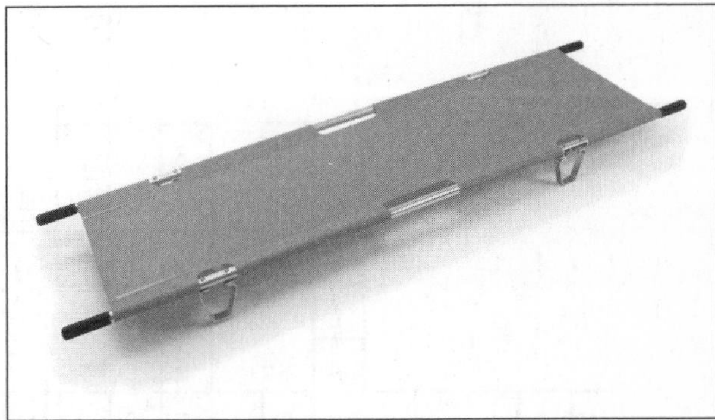

B. Stretcher (page 272)

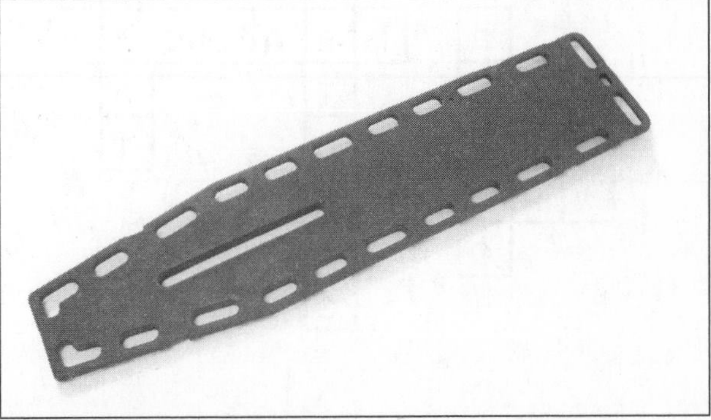

C. Spine board (page 272)

Vocabulary

1. **Rescue mode:** Those activities directed at locating endangered persons at an emergency incident, removing those persons from danger, treating injured victims, and providing transport to an appropriate healthcare facility. (page 269)
2. **Backup team:** Individuals who function as a stand-by rescue crew of relief for those entering the hot zone (entry team). Also referred to as backup personnel. (page 266)
3. **Triage:** The process of sorting victims based on the severity of their injuries and medical needs to establish treatment and transportation priorities. (page 267)

Fill-in

1. Nonambulatory (page 267)
2. rescue (page 268)
3. six (page 269)
4. two-person extremity carry (page 276)
5. clothes drag (page 279)
6. webbing (page 282)
7. long backboard (page 284)
8. two (page 266)
9. triage (page 267)
10. necessary (page 266)

True/False

1. T (page 266)
2. T (page 269)
3. F (page 269)
4. F (page 269)
5. T (page 269)
6. T (page 267)
7. T (page 278)
8. F (page 284)
9. T (page 270)
10. F (page 274)

Short Answer

1. Emergency drags include the clothes drag, blanket drag, standing drag, webbing sling drag, fire fighter drag, and emergency drag from a vehicle. (page 278)
2. Simple Triage and Rapid Treatment Triage (START) is a system that prompts the responder to assess a patient's breathing rate, pulse rate, and mental status, so as to assign a treatment priority to the victim. (page 268)

Word Fun

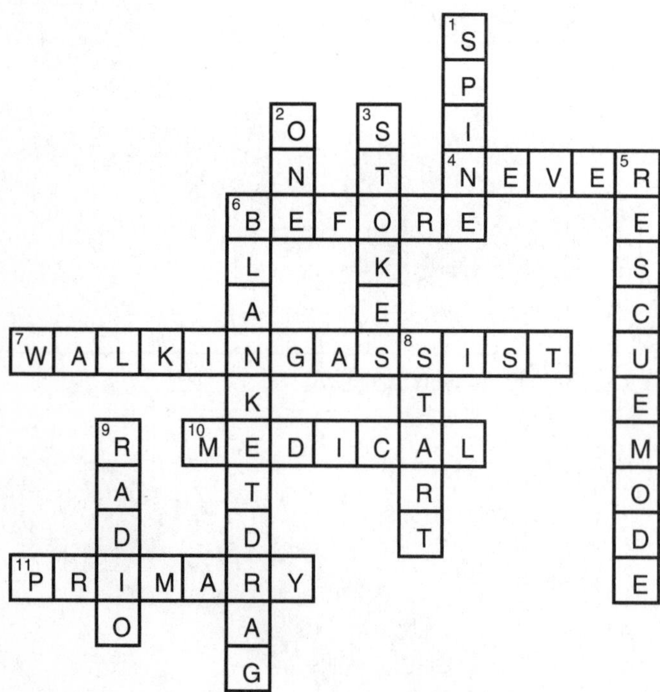

Fire Alarms

1. When faced with a potential victim rescue, emergency responders should first ensure that enough responders are on the scene to make the attempt. At least five responders will be needed. Two for the entry team, two for the back-up team, and one to staff an emergency decontamination process. You will want to work in teams of two, use SCBA and structural fire protective clothing (since chemical-protective clothing is not yet available), and approach the scene from upwind. (page 266)
2. Remove her from the immediate area to an area upwind and begin resuscitation immediately. As always, there are no absolutes when it comes to delivery of medical care to save a life. In some extreme or unusual cases, medical care may need to be delivered in the warm or hot zone prior to or concurrent with decontamination. In those situations, hazardous materials responders and medical personnel must balance the need for performing life-saving interventions with the need for ensuring decontamination. This decision is made on a case-by-case basis, with considerations being given to the nature and severity of the incident, the medical needs of the patients, and the need to perform decontamination prior to rendering care. (page 269)

Skill Drills

Skill Drill 12-1: Performing a One-Person Walking Assist

1. Help the victim to **stand**.
2. Have the victim place his or her arm around your neck, and hold onto the victim's **wrist**, which should be draped over your shoulder. Put your free arm around the victim's **waist** and help the victim to walk.

Skill Drill 12-2: Performing a Two-Person Walking Assist

1. Two responders stand facing the victim, one on each side of the victim.

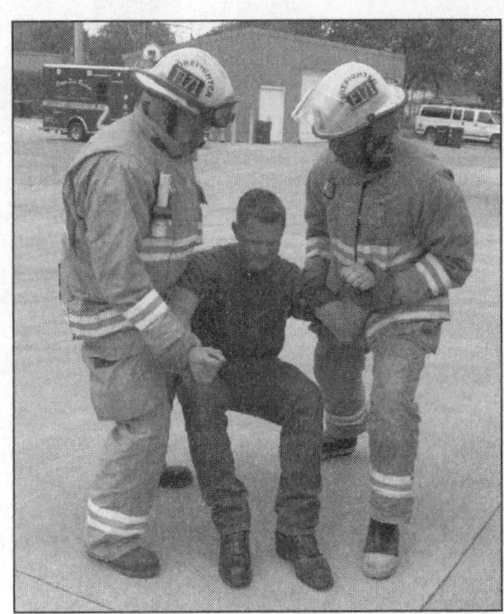

2. The responders assist the victim to a standing position.

3. Once the victim is fully upright, drape the victim's arms around the necks and over the shoulders of the responders, each of whom holds one of the victim's wrists.

4. Both responders put their free arm around the victim's waist, grasping each other's wrists for support and locking their arms together behind the victim.

5. Assist walking at the victim's speed.

Skill Drill 12-3: Performing a Two-Person Extremity Carry

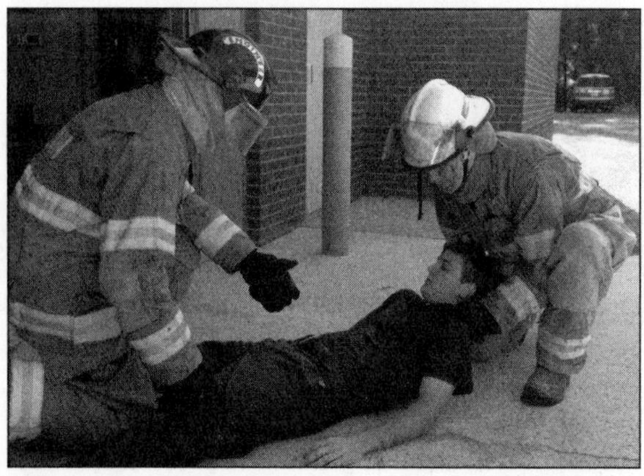

1. Two responders help the victim to sit up.

2. The first responder kneels behind the victim, reaches under the victim's arms, and grasps the victim's wrists.

3. The second responder backs in between the victim's legs, reaches around, and grasps the victim behind the knees.

4. The first responder gives the command to stand and carry the victim away, walking straight ahead. Both responders must coordinate their movements.

Skill Drill 12-4: Performing a Two-Person Seat Carry

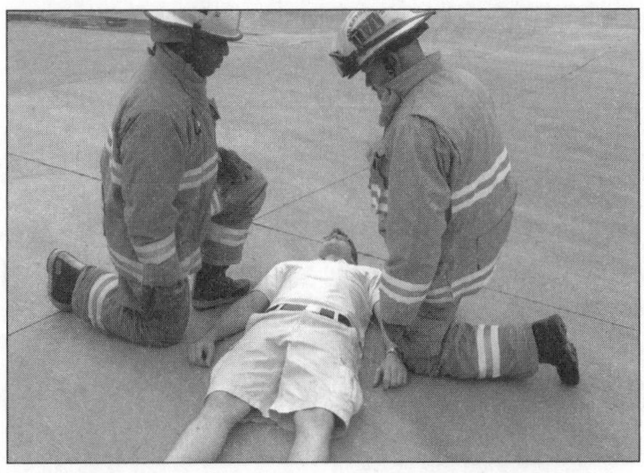

1. Kneel beside the victim near the victim's hips.

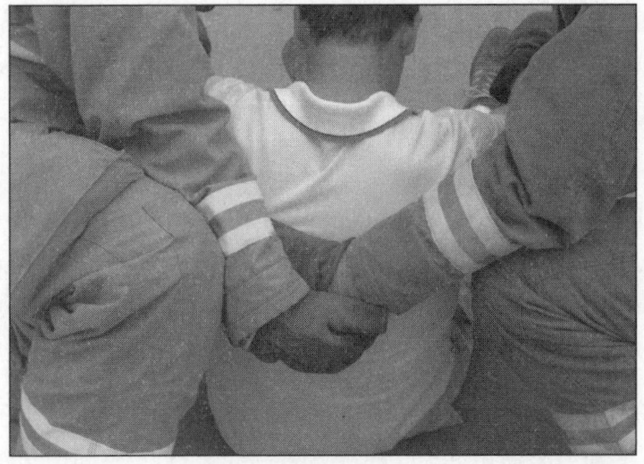

2. Raise the victim to a sitting position and link arms behind the victim's back.

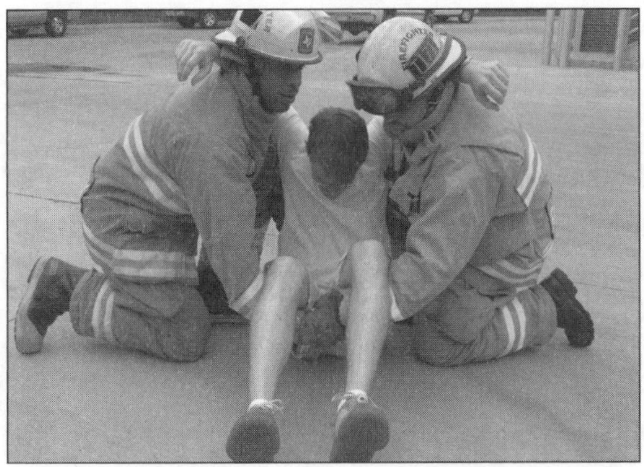

3. Place your free arms under the victim's knees and link arms.

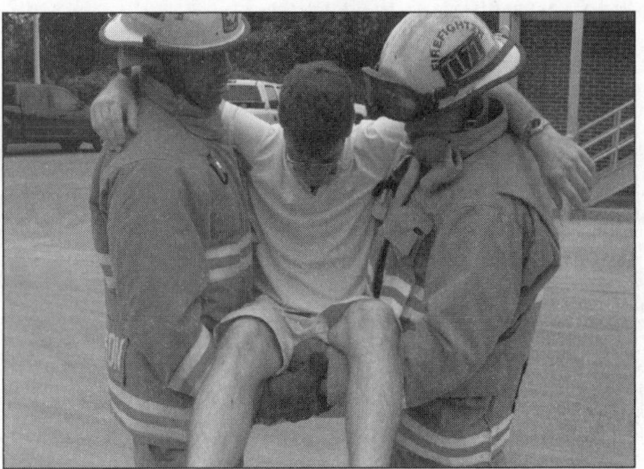

4. If possible, the victim puts his or her arms around your necks for additional support.

Skill Drill 12-5: Performing a Two-Person Chair Carry

1. One responder stands behind the seated victim, reaches down, and grasps the **back** of the chair.
2. The responder tilts the chair slightly backward on its rear legs so that the second responder can step between the legs of the chair and grasp the tips of the chair's front legs. The victim's **legs** should be **between** the legs of the chair.
3. When both responders are correctly positioned, the responder behind the chair gives the command to **lift** and walk away. Because the chair carry may force the victim's **head** forward, watch the victim for **airway** problems.

Skill Drill 12-6: Performing a Cradle-in-Arms Carry

1. Kneel beside the child, and place one arm around the child's **back** and the other arm under the **thighs**.
2. Lift slightly and roll the child into the hollow formed by your **arms** and **chest**.
3. Be sure to use your **leg** muscles to stand.

Skill Drill 12-7: Performing a Clothes Drag

1. Crouch behind the victim's head, and grab the shirt or jacket around the collar and **shoulder** area.
2. Lift with your **legs** until you are fully upright. Walk **backward**, dragging the victim to safety.

Skill Drill 12-8: Performing a Blanket Drag

1. Stretch out the **material** you are using next to the victim.
2. Roll the victim onto one side. Neatly bunch one **third** of the material against the victim's body.
3. Lay the victim back down (**supine**). Pull the bunched material out from underneath the victim and **wrap** it around the victim.
4. Grab the material at the **head** and drag the victim **backward** to safety.

Skill Drill 12-9: Performing a Standing Drag

1. Kneel at the head of the **supine** victim.
2. Raise the victim's head and torso by **90** degrees, so that the victim is leaning against you.
3. Reach under the victim's **arms**, wrap your arms around the victim's **chest**, and lock your arms.
4. Stand straight up using your legs. **Drag** the victim out.

Skill Drill 12-10: Performing a Webbing Sling Drag

1. Place the victim in the **center** of the loop so the webbing is behind the victim's **back**.
2. Take the **large** loop over the victim and place it above the victim's head. Reach through, grab the webbing behind the victim's back, and pull through all the excess webbing. This creates a loop at the top of the victim's head and **two** loops around the victim's arms.
3. Adjust your hand placement to protect the victim's **head** while dragging.

Skill Drill 12-11: Performing a Fire Fighter Drag

1. Tie the victim's wrists together with **anything** that is handy.
2. Get down on your hands and knees and **straddle** the victim.
3. Pass the victim's tied hands around your neck, **straighten** your arms, and drag the victim across the floor by crawling on your hands and knees.

Skill Drill 12-12: Performing a One-Person Emergency Drag from a Vehicle

1. Grasp the victim under the arms and **cradle** his or her head between your arms.
2. Gently **pull** the victim out of the vehicle.
3. Lower the victim down into a **horizontal** position in a safe place.

Skill Drill 12-13: Performing a Long Backboard Rescue

1. The first responder provides in-line manual support of the victim's head and cervical spine.

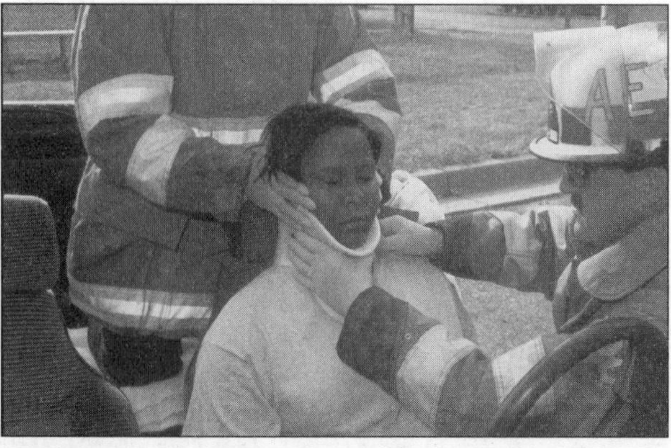

2. The second responder gives commands and applies a cervical collar.

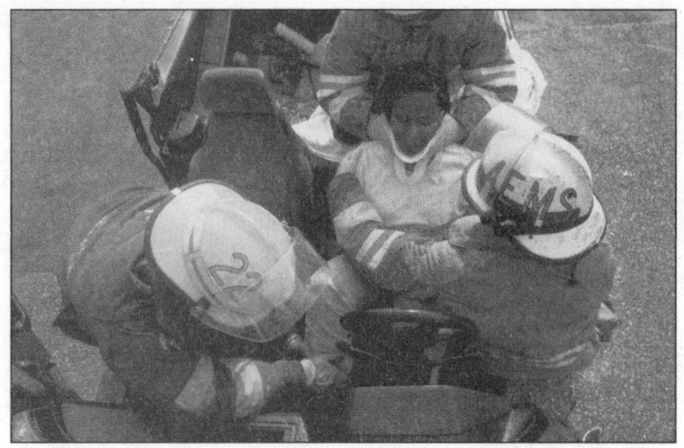

3. The third responder frees the victim's legs from the pedals and moves the legs together without moving the victim's pelvis or spine.

4. The second and third responders rotate the victim as a unit in several short, coordinated moves. The first responder (relieved by the fourth responder as needed) supports the victim's head and neck during rotation (and later steps).

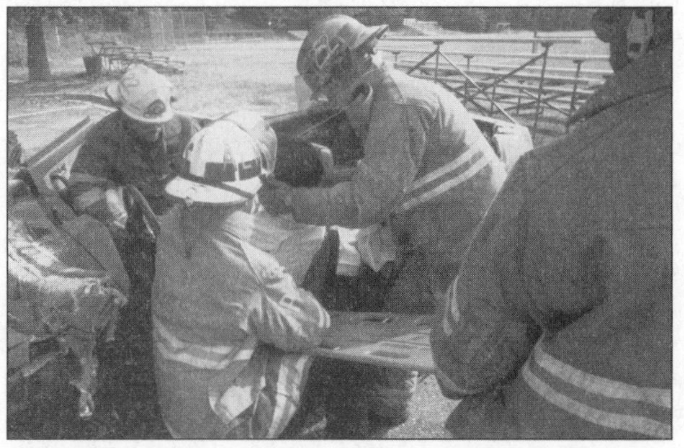

5. The first (or fourth) responder places the backboard on the seat against the victim's buttocks. The second and third responders lower the victim onto the long backboard.

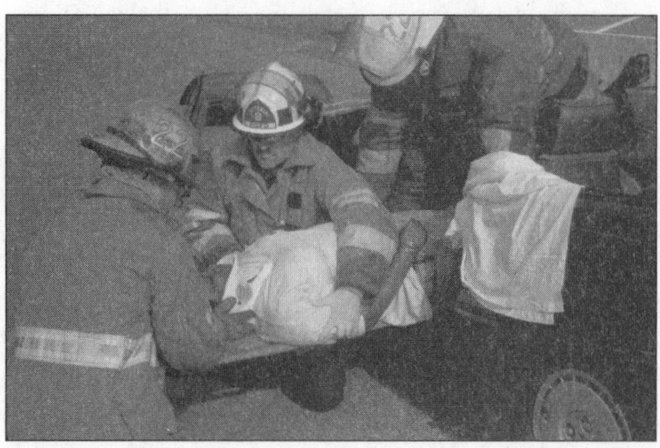

6. The third responder moves to an effective position for sliding the victim. The second and third responders slide the victim along the backboard in coordinated, 8" to 12" (20-cm to 31-cm) moves until the victim's hips rest on the backboard.

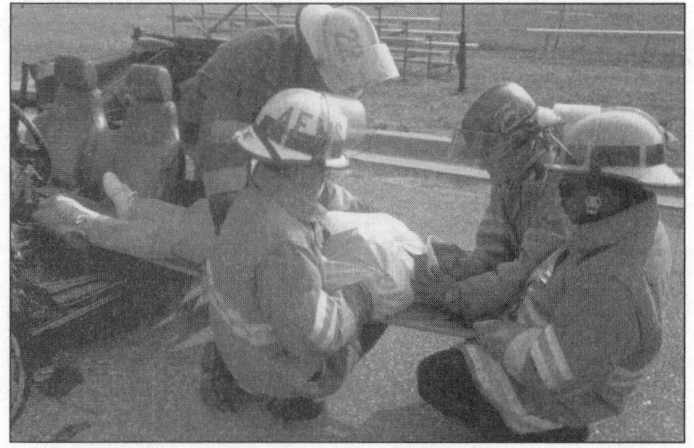

7. The third responder exits the vehicle and moves to the backboard opposite the second responder. Working together, they continue to slide the victim until the victim is fully on the backboard.

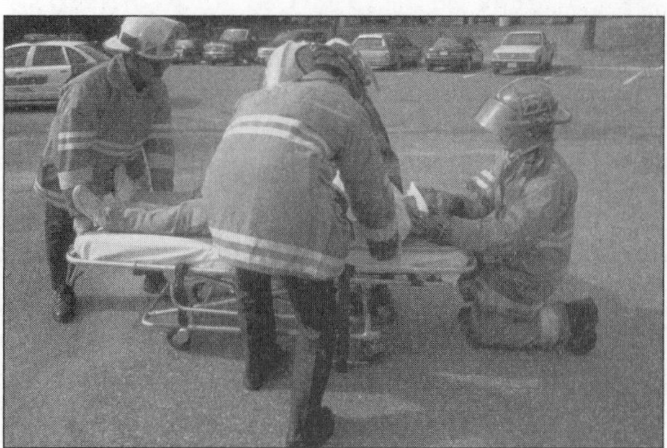

8. The first (or fourth) responder continues to stabilize the victim's head and neck, while the second, third, and fourth responders carry the victim away from the vehicle.

Chapter 13: Mission-Specific Competencies: Response to Illicit Laboratories

Matching

1. E (page 296)
2. A (page 297)
3. B (page 297)
4. C (page 297)
5. D (page 294)
6. H (page 297)
7. F (page 297)
8. G (page 297)
9. J (page 300)
10. I (page 297)

Multiple Choice

1. B (page 297)
2. C (page 297)
3. D (page 297)
4. C (page 298)
5. D (page 297)

Vocabulary

1. **Clandestine drug laboratory:** An illicit operation consisting of a sufficient combination of apparatus and chemicals that either has been or could be used in the manufacture or synthesis of controlled substances. (page 295)
2. **Explosive ordnance disposal (EOD) personnel:** Personnel trained to detect, identify, evaluate, render safe, recover, and dispose of unexploded explosive devices. (page 297)
3. **Illicit laboratory:** Any unlicensed or illegal structure, vehicle, facility, or physical location that may be used to manufacture, process, culture, or synthesize an illegal drug, hazardous material/WMD device, or agent. (page 294)

Fill-in

1. methamphetamine (page 296)
2. rescue rope (page 297)
3. 72 hours (page 299)
4. Drug Enforcement Administration (page 299)
5. explosive ordnance disposal (page 297)

True/False

1. T (page 296)
2. T (page 294)
3. T (page 295)
4. T (page 296)
5. F (page 295)
6. T (page 297)
7. F (page 297)
8. F (page 296)
9. T (page 301)
10. T (page 301)

Short Answer

1. Methamphetamine chemicals and their legitimate uses include: (1) Anhydrous ammonia (fertilizer), (2) Methanol (gasoline additive), (3) Ether (engine starter), (4) Toluene (brake cleaner), (5) Sulfuric acid (brake cleaner), (6) Muriatic acid (pool supply), (7) Iodine (medical use), (8) Kerosene (camp stove fuel), (9) Acetone (paint remover), (10) Lithium (batteries), (11) Sodium hydroxide (lye, drain cleaner), and (12) Red phosphorous (matches, flare igniters). (page 296)

Word Fun

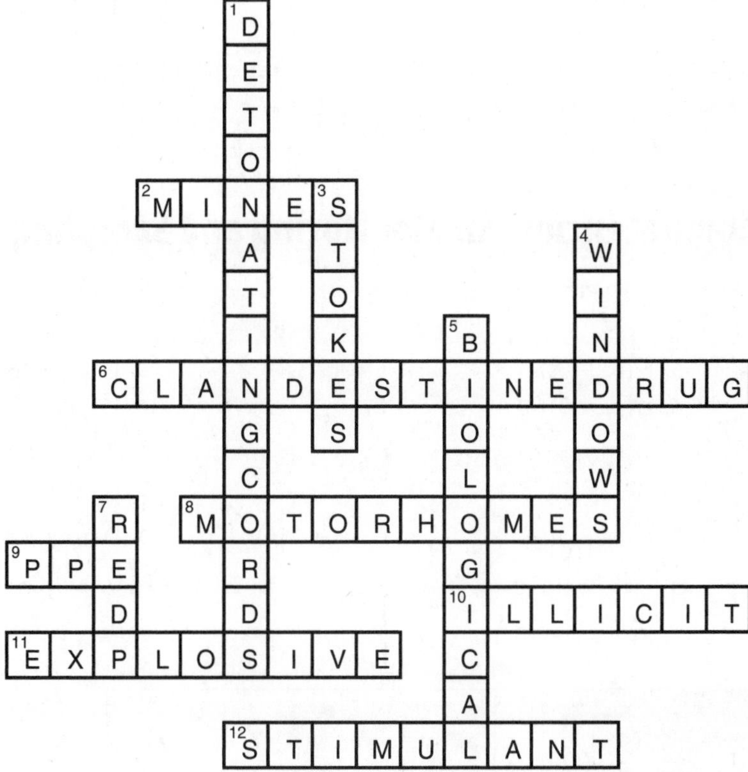

Fire Alarms

1. These are indicators that explosive materials or devices may be present. Your immediate actions must be to leave the area right away and notify the appropriate explosive ordnance disposal (EOD) personnel. (page 297)
2. Provide clear instructions to the law enforcement officers entering the decontamination area. Realize that while they may be trained to wear PPE, law enforcement officers may not be trained in decontamination procedures. Instruct them to point any weapons in a safe direction, unload the weapons completely, lock back the firing mechanism, and engage the safety selector switch.

   Consult with canine officers to determine the best way to handle the animal if decontamination is required. Instruct law enforcement officers handling a canine to maintain control of the animal during the entire process. Have them apply a muzzle to the animal if necessary.

   Instruct law enforcement officers handling prisoners to maintain control of the prisoners at all times. It is recommended that each prisoner be controlled by two law enforcement officers during this process, so that the officers can be decontaminated as well. Begin decontamination procedures as necessary. (page 301)

Skill Drills

Skill Drill 13-1: Identifying and/or Avoiding Potential Unique Safety Hazards

1. Visually assess the structure or property that is suspected to contain a **laboratory** operation for outward warning signs, such as the presence of security and **surveillance** systems (including triggering devices and booby traps), precursor **chemical** containers, laboratory equipment, or hostile **occupants**.
2. Establish a safe **containment** **perimeter** based on the hazards identified.
3. Notify the appropriate **law** **enforcement** personnel, technicians, and allied professionals based on the hazards identified.
4. Make an assessment of any **victims** who may be present and any **symptoms** they are presenting.

Skill Drill 13-2: Conducting Joint Hazardous Materials/EOD Operations

1. Discuss with law enforcement or EOD personnel those **materials** or **devices** that are potentially explosive and/or hazardous.
2. Develop a joint **response** plan, if necessary, to render the device or materials safe for collection as **evidence**.
3. Develop a **decontamination** plan to support EOD personnel and equipment.

Chapter 14: Mission-Specific Competencies: Air Monitoring and Sampling

Matching

1. B (page 314)
2. A (page 310)
3. D (page 313)
4. C (page 315)
5. F (page 315)
6. E (page 315)
7. I (page 320)
8. G (page 320)
9. J (page 316)
10. H (page 310)

Multiple Choice

1. B (page 310)
2. B (page 316)
3. B (page 316)
4. C (page 314)
5. D (page 316)

Labeling

1. Types of detectors and monitors.

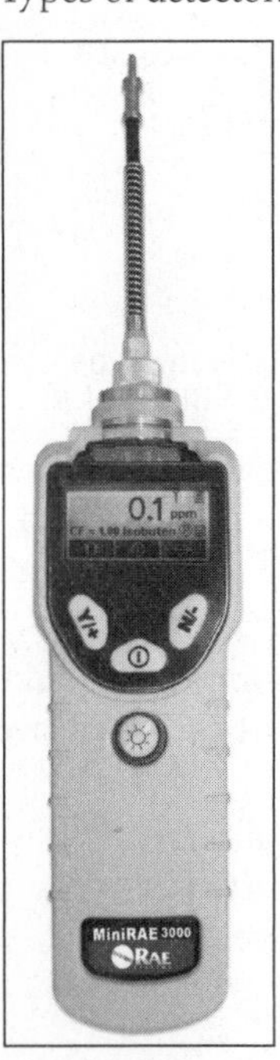

A. Photo-ionization device (PID)

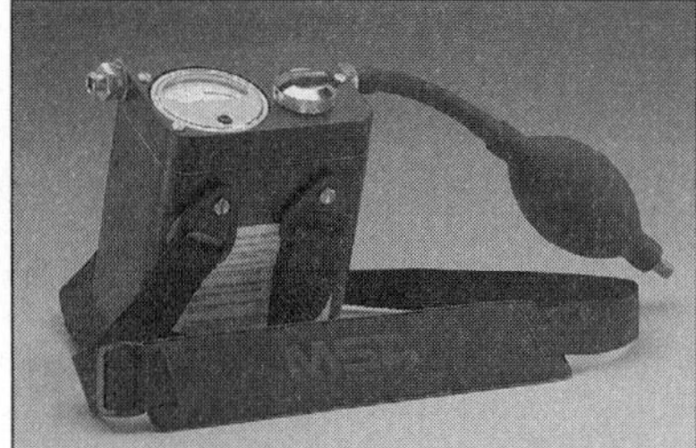

B. MSA2a combustible gas indicator

C. Personal dosimeters

Vocabulary

1. **Volatile organic compound:** Any organic (carbon-containing) compound that is capable of vaporizing into the atmosphere under normal environmental conditions. (pages 308, 322)
2. **Relative response factor:** A curve that accounts for the different types of gases and vapors that might be encountered other than the one used for calibration of a flammable gas detector and/or monitor. (page 311)
3. **Situational awareness (SA):** A term primarily used in the military, especially the Air Force, that describes the process of observing and understanding the visual cues available, orienting yourself to those inputs relative to your current situation, and making rapid decisions based on those inputs. (page 309)

Fill-in

1. Calibration (page 309)
2. Situational awareness (page 309)
3. relative response factor (page 311)
4. 20.9 (page 315)
5. pH paper (page 318)
6. zeroed (page 310)
7. recovery time (page 310)
8. 35 (page 316)
9. Chemical test strips (page 319)
10. Multi-gas meters (page 316)

True/False

1. T (page 307)
2. T (page 315)
3. F (page 314)
4. T (page 317)
5. T (page 317)
6. F (page 319)
7. F (page 317)
8. T (page 320)
9. T (page 320)
10. F (page 310)

Short Answer

1. The 10 basic actions for detection and monitoring as identified in NFPA 472:

   (1) Attempt to identify the source and nature of potential contamination (prior to entry).

   (2) Research and understand the nature of any identified atmospheric contamination. Do not assume that only one hazard exists.

   (3) Select the proper personal protective equipment for the task.

   (4) Select the appropriate instrument(s) for the task.

   (5) Properly prepare the instrument(s) for use.

   (6) Prioritize your monitoring areas.

   (7) Develop an overall monitoring plan.

   (8) Confirm all readings when obtained and record when appropriate.

   (9) Establish action levels (readings from the devices) that could dictate tactics or other actions.

   (10) At the conclusion of the incident, survey the areas again to confirm that the hazard has been mitigated. (page 309)

Word Fun

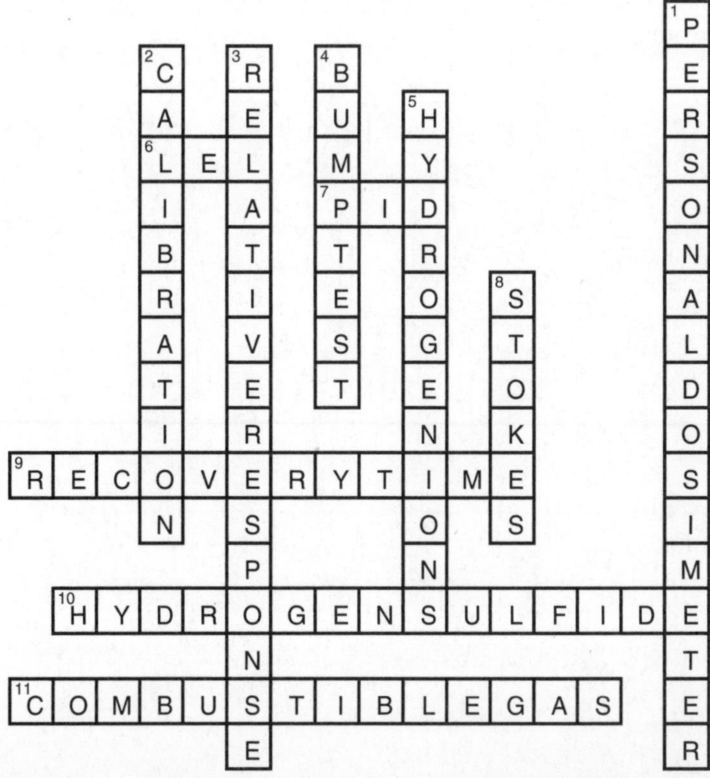

Fire Alarms

1. The risk of explosion is significant because the product has a very low flash point, and the vapors somewhere at the scene may be in the LEL–UEL range, especially in the absence of a ventilation system. Most combustible gas indicators are set to alarm at a value of 10% of the substance for which they have been calibrated. The highest level of danger occurs when the atmosphere reaches 100% of the LEL. At this point, the operator should interpret the reading as indicating that a sufficient amount of flammable gas or vapor is present in the air to support combustion. (page 314)
2. The student's exposure exceeds the STEL, and if the student was exposed for 8 hours at the stated rate, then his exposure would exceed the TWA REL. The student experienced the symptoms described due to both exposure and contamination. (page 317)

Skill Drills

Skill Drill 14-3: Using a Multi-Gas Meter to Provide Atmospheric Monitoring (after the proper level of PPE is selected)

1. Understand the manufacturer's recommendations and local standard operating procedures for **multi-gas meter** use. Turn on the device and zero it in a clean atmospheric environment. Let the device warm up. Perform a **bump test**.
2. Approach the hazardous material and **monitor** the atmosphere.
3. Interpret the meaning of the **readings**. Return to a safe atmosphere. Return the meter to zero and follow the appropriate procedures to turn the meter off and return the meter to **service**.

Photo Credits

Page 21, Part A and B Courtesy of Polar Tank Trailer L.L.C.
Page 21, Part C Courtesy of National Tank Truck Carriers Association
Page 21, Part D Courtesy of Rob Schnepp
Page 22, Part E and F Courtesy of Jack B. Kelly, Inc.
Page 22, Part G Courtesy of Polar Tank Trailer L.L.C.
Page 22, Part H Courtesy of private source
Page 30, Part 3 Courtesy of The DuPont Company
Page 46, Part A Courtesy of Dr. Saeed Keshavarz/RCCI (Research Center of Chemical Injuries)/IRAN
Page 46, Part C Courtesy of CDC
Page 54, Part A Image © Lakeland Industries Inc. All rights reserved.
Page 54, Part B © Photodisc
Page 65, Skill Drill 7-7 Courtesy of Rob Schnepp
Page 99, Part A and B Courtesy of Ferno-Washington, Inc. www.ferno.com
Page 100, Part C Courtesy of Ferno-Washington, Inc. www.ferno.com
Page 121, Part A Courtesy of RAE Systems
Page 121, Part B Courtesy of MSA – The Safety Company
Page 121, Part C Courtesy of S.E. International, Inc. www.seintl.com
Page 133, Part A and B Courtesy of Polar Tank Trailer L.L.C.
Page 133, Part C Courtesy of National Tank Truck Carriers Association
Page 133, Part D Courtesy of Rob Schnepp
Page 133, Part E and F Courtesy of Jack B. Kelly, Inc.
Page 133, Part G Courtesy of Polar Tank Trailer L.L.C.
Page 133, Part H Courtesy of private source
Page 140, Part A Courtesy of Dr. Saeed Keshavarz/RCCI (Research Center of Chemical Injuries)/IRAN
Page 140, Part C Courtesy of CDC
Page 143, Part A Image © Lakeland Industries Inc. All rights reserved.
Page 143, Part B © Photodisc
Page 159, Part A, B, and C Courtesy of Ferno-Washington, Inc. www.ferno.com
Page 168, Part A Courtesy of RAE Systems
Page 168, Part B Courtesy of MSA – The Safety Company
Page 168, Part C Courtesy of S.E. International, Inc. www.seintl.com

Unless otherwise indicated, all photographs and illustrations are under copyright of Jones and Bartlett Publishers, LLC, and were photographed by Glen E. Ellman or were a courtesy of the Maryland Institute for Emergency Medical Services Systems.